Saving the Sacred Sea

Saving the Sacred Sea

THE POWER OF CIVIL SOCIETY IN AN AGE OF
AUTHORITARIANISM AND GLOBALIZATION

KATE PRIDE BROWN

OXFORD
UNIVERSITY PRESS

OXFORD
UNIVERSITY PRESS

Oxford University Press is a department of the University of Oxford. It furthers
the University's objective of excellence in research, scholarship, and education
by publishing worldwide. Oxford is a registered trade mark of Oxford University
Press in the UK and certain other countries.

Published in the United States of America by Oxford University Press
198 Madison Avenue, New York, NY 10016, United States of America.

© Oxford University Press 2018

CIP data is on file at the Library of Congress
ISBN 978-0-19-066095-6 (pbk.)
ISBN 978-0-19-066094-9 (hbk.)

9 8 7 6 5 4 3 2 1

Paperback printed by Webcom, Inc., Canada
Hardback printed by Bridgeport National Bindery, Inc., United States of America

For

Mr. C

Contents

Acknowledgments

This book would not have been possible without the assistance and support of many individuals and organizations. Any attempt to produce a full accounting of them here would undoubtedly fail, but some key individuals and groups are especially worthy of thanks.

First and foremost, I owe an enormous debt of gratitude to the community of environmental activists in Irkutsk and around Lake Baikal. Their companionship was so warm that I barely noticed my Siberian winter. I came away with enormous admiration and affection for the Baikal region and for the many people who call it home. I want to thank and acknowledge especially the individuals who comprise Baikal Environmental Wave, the Tahoe-Baikal Institute, the Great Baikal Trail, Reviving Siberian Land, Defending Baikal Together, and Let's Make Irkutsk Eco-Logical. These organizations are full of people who helped my fieldwork in a variety of ways, but a few are of particular note. I want to especially thank Anya Belova, Elena Chubakova, Evgeniy Maryasov, Marina Rikhvanova, Andrei Suknev, Jennie Sutton, and Elena Tvorogova. There are many others who deserve acknowledgment and thanks, and many of them appear in the pages of this book under pseudonyms to protect their privacy. Hopefully they know who they are and that they have my deep gratitude.

In the United States, Bob Birkby, Sarah Buck, Gary Cook, Ariadna Reida, Bud Sheble, and Jennifer Smith-Lee were invaluable resources, as well as great folks. Scott and Catherine Cecci were superb hosts during my stay at Lake Tahoe. Bob Harris, who passed away in 2013, was helpful, kind, and a great enthusiast for Russia and Lake Baikal. I wish we could have talked more.

My fieldwork in Irkutsk was funded by a US Student Fulbright Fellowship from the Institute for International Education, and I received additional support from the Horowitz Foundation for Social Policy during the writing process. My institutional affiliation in Russia was with the Irkutsk State

Linguistic University (ISLU). The staff of the foreign relations department was incredibly helpful in facilitating my research needs. In a country known for its stultifying red tape, they smoothly and reliably resolved my every issue. Maria Ivanova in the history department was very helpful as I sought to navigate the library and archive system. The librarians and archivists at the Molchanov-Sibirskii Irkutsk Regional State Research Library and the Irkutsk Regional Archive of Recent History were reliable, resourceful, and often very friendly. I benefited from the research assistance of a small corps of individuals who helped me gather scores of archived newspaper articles— I owe thanks to all of them.

This book began as a dissertation project that I completed at Vanderbilt University, where I received excellent mentorship from stellar minds. In addition to being my advisor, Richard Lloyd was a kindred spirit. He never sought to rein in my theoretical reach, as long as it did not exceed my empirical grasp. George Becker, David Hess, Larry Isaac, and Frank Wcislo also played formative roles in my thinking about this project, and I was honored to have their mentorship. Laura Henry has been a wonderful colleague to me from my earliest days of graduate school. She read the entire manuscript closely and carefully, and her detailed notes rendered the finished product much improved. The project would never have made it off the ground without the initial help and mentorship of Bob Desjarlais, Melissa Frazier, and Shahnaz Rouse. Finally, I thank my colleagues in the School of History and Sociology at Georgia Tech who have been hugely supportive and gave me plenty of time to finish the last leg of writing. Sofia Slutskaya in the Georgia Tech library helped me find several important citations.

I would like to thank Oxford University Press for publishing this book, and most especially to Angela Chnapko. I could not have asked for a better editor. I also thank Heather Hambleton, who provided copyediting, and anonymous reviewers for their comments. From start to finish I have been assisted in myriad ways. Any remaining errors are my own.

We all stand on the shoulders of giants, and it is through the inspiration of great teachers that we have the audacity to climb to that summit. My path has been made smoother and much more interesting because of the accomplished educators that surround me. My parents, Richard and Dana Pride, ensured that the life of the mind imbued our house when I was growing up, and our family dinner conversations were second to none. My brother, Ethan, is an intellectual inspiration, and my sister, Sarah, beautifully melded creativity with learning. My husband, Bill, continuously broadens my horizons; indeed, he has opened whole new worlds for me. My daughter, Scheherazade, teaches me more and more every day—ineffable lessons about life and humanity that I might never have learned but for her.

We all have teachers, but only very occasionally is there one who makes so dramatic and lasting imprint that the lessons never fade into the background but stay close and tight to the surface of one's mind across the decades. Seldom do we look back on our primary education and point to a single year that became a turning point in one's conception of what schooling is and what learning ought to be. In the fifth grade, I was beyond fortunate to find myself seated in the classroom of Gary Christy—Mr. C—a man who did not think it was presumptuous to teach critical thought to ten year olds. I embarked on a year of marvels, enraptured by the possibilities of creativity, new ideas, and intellectual challenge. I was so grateful to him for this year, which stood out as markedly different from any education I had received before, that I found myself wanting to do something remarkable with my life in order to better honor him. I remember thinking, way back then, that if I ever wrote a book, I would dedicate it to Mr. Christy. Because of the importance of what he did as a teacher, for me and for countless other children across the years, I thank him. This one is for you, Mr. C.

Saving the Sacred Sea

1

Introduction

A bonfire blazed and crackled, casting its orange glow on the twenty-plus faces clustered around it. The heat from the flames warmed our fronts, while the chilly Siberian night threatened us from behind. Although it was mid-August, birch trees were already beginning to yellow, and we would wake in the morning to find a crust of ice over any standing water. Summer was ending, and autumn would soon begin. The smell of frost mingled with wood smoke on this night, our last in the *taiga*, the boreal forest that covers much of Siberia.

Although it was cold and late, no one was willing to leave the fire, as though the night and our time together might not end if only we kept the blaze burning and the guitar strumming. We had come together in this place as strangers only two weeks before, but since then had become fast friends.

Now we sat together around the fire, on benches that we had made ourselves, watching as the remainder of our wood was consumed in the conflagration. An invisible thread seemed to bind us together that night, and we communicated freely—English, Russian, gesture, facial expression—the myriad ways we had learned to read one another over the last two weeks.

The night crept on, and the songsters were beginning to exhaust their repertoires. Eyelids were drooping, and the woodpile was low. There was a pause in the joviality as each sat quiet.

"It's midnight," Olga said.[1]

"What time exactly?" Larisa asked her.

"Exactly midnight," Olga answered. "It just turned."

"We should sing the national anthem," Larisa replied, half-joking. She was referencing Radio Russia, the principle radio station in the country, which goes off the air at midnight and ends its broadcast with the national anthem.

"Yes, we should!" agreed Polya, and soon they were rousing the group to stand and sing. We all obliged; the twenty of us, standing around the campfire, began a heartfelt variation of "*Rossiya—svyashchennaya nasha derzhava...,*"[2] a multitude of voices carrying out into the night. Our non-Russian speaking companions stood and smiled in appreciation and support. As the final strains of the hymn echoed in the tall pines, we laughed and seated ourselves again.

"Hey, we should sing *all* our national anthems," Olga suggested.

"Yeah, what does your anthem sound like?" Larisa asked the brother and sister from Portugal.

"Sing it!" several others urged them on.

They began a very enthusiastic (albeit not always in key) rendition of the Portuguese national anthem. Next was the German national anthem, sung enchantingly by Silke. I followed with the "Star-Spangled Banner," mustering more patriotism than I knew I possessed. Thomas and Emilie stole the show with the merry opening of "La Marseillaise," with everyone clapping along. Antonio sang the Italian national anthem, which none of us knew, but all were glad to hear.

"Now for the British!" Vadim said.

Peter and Jane exchanged glances. "I don't know it, do you?" they both muttered between themselves, outing their lack of interest in the monarchy.

"Oh, we all sang, you *have* to sing it," and similar lines of encouragement came out of the crowd. Obligingly, they managed between the two of them to sing the first two lines of "God Save the Queen," and then hummed the rest.

"Is that all?" Larisa asked.

"I think so," Katya answered.

There was a pause.

"Do you know the Buryat national anthem?" Polya asked Roma, a young Buryat man from the Irkutsk *oblast* [province]. Roma shook his head.

"I do," said a small voice. We all looked, surprised, at Fedya, a dark, quiet man from Moscow, who seldom spoke.

"Well . . . sing it!" Olga said. There was a pause, and then Fedya's thin voice carried forth a hauntingly beautiful hymn to the Republic of Buryatia, an ethnic state within the Russian Federation, whose people were indigenous to Lake Baikal. The slow, lyrical minor key was a sedate contrast to such rousing hymns as "La Marseillaise" and the "Star-Spangled Banner." Roma listened to his people's song, coming from the mouth of a quiet Muscovite, with his eyes cast down into the fire. He looked pleased.

The twenty-three people around this fire were all volunteers with the Great Baikal Trail project, a volunteer-based nonprofit organization in Irkutsk, Russia, whose goal is to promote sustainable eco-tourism by building a network of hiking trails around Lake Baikal. Recruits from around the world participate in two-week "volunteer vacations" to help build and maintain low-impact wilderness trails. They work collectively to put into place a particular, shared vision for the human-environment relationship at Lake Baikal. They comprise an environmental civil society.

This one microcosm of environmental civil society acts out its mission in a particular place—a five-kilometer stretch of a remote Siberian forest. Yet it drew support for this activity from an array of sympathetic individuals who, in turn, reside thousands of miles away. As the above anecdote suggests, participants receive the bounty of cultural exchange: they form bonds of friendship, breaking down prejudicial barriers and expanding the "imagined community" (Anderson 1983) to peoples worldwide.

But despite the surprising assortment of national anthems sung out that night in the Siberian *taiga*, it is worth noting that the foreign volunteers were uniformly Western. While the Great Baikal Trail draws strength from its international reach, those foreigners who are best positioned to help come from a select geographic area. Thus, this little party of volunteers singing in the woods also represents the geographic imbalance of transnational activism. Local environmental civil society may benefit from international cooperation, but the global space it draws upon is itself beset by inequities and power imbalances that necessarily carry forward into these local efforts at environmental protection.

Furthermore, the activity in which these volunteers engage is itself a response to another social vector: global capitalism. There are many different forms of economic development, and locations such as Lake Baikal respond to the push and pull of the global marketplace in goods and services. The international team of volunteers on the Great Baikal Trail project is driven by one particular form of development. Environmental protection via tourism is a response to the alternative—and dominant—model of development in the region: heavy industry and natural resource extraction. Tourism is also a major global industry, and it is one that the Great Baikal Trail seeks to channel into its own backyard as sustainably as possible.

But despite the globalized economy and the transnational environmental ethic that brought these foreign volunteers to Siberia, in order to participate, they all had to navigate the bureaucratic rigmarole of the Russian state's visa regime. More than a hassle, it was a reminder that the nation-state has not vanished in the "globalized" twenty-first century: the world is not flat. Not only are political boundaries solidly present, they contain within their sovereign territories a host of different political governance systems, whose impact on the lives of their subjects may be more or less benign. On the trip, there were occasionally tentative, prodding questions between the Russian volunteers and their foreign guests: *Do you like your President Obama? Did you know that the protests in Moscow were not shown on TV? What do you think about Syria?* In 2012, Vladimir Putin had just returned to the Russian presidency, and the significance of his return was still in the realm of supposition. But the understanding that national political powers mediated our temporarily conjoined lives was nonetheless palpable.

There are many narratives that run through the story of these twenty-three people, singing national anthems on a cold Siberian night. Their presence in those woods, the many countries from which they hail, the international tourism industry that brought them there, the red-earth switchback trail they cooperatively built in order to counteract a global market in raw materials, the authoritarian government of the country that engulfed them, its long history of centralized domination: these threads form a tapestry of social forces at work, operating at many different spatial scales, from the very local to the truly global. While countless influences shape the campfire performance that took place that night, these twenty-three people have come together of their own volition, sharing an abiding concern for Lake Baikal; and together, for at least five kilometers, they changed the world. They embody an environmental civil society.

Siberia is not the place that most people go to study civil society. To the Western mind, it heralds a land that is cold, barren, desolate, and remote. It brings forth images of the *gulag*—the ugliest of Soviet history and the very emblem of totalitarianism. Given these customary associations, the average reader may be inclined to ask: What could Siberia possibly have to teach us about civil society? I would answer: everything.

Siberia is the home to Lake Baikal, the oldest, deepest, and most voluminous lake on Earth. It is called the Sacred Sea. It is also home to a strong environmentalist community that has worked for decades to protect this treasured lake. Its activism began under the repressive Soviet government and continues through this day, in a civil society that spans eras. Baikal's activist community is also a space of interconnectivity, a meeting place of East and West. Despite its remote location, Baikal is a hotbed of international attention. Environmental activism around Lake Baikal is so bound up with transnationality it is impossible to parse the purely local from the global. It is precisely here at Lake Baikal, in this place—so removed from the Western ideal type of a robust civil society—that we may uncover new understandings of what civil society is and what it does.

Modern theories of civil society grew out of grassroots struggles against authoritarianism in the Soviet Union and elsewhere around the globe in the late 1980s. The civil society concept that this literature portrayed and produced inevitably described the democratic West. It was a theory born of comparison: What does the West possess that allows its people to curb state power while Soviet citizens cannot? The statement of the problem determined its answer. Ever since, the Western model of civil society has become *the* model of civil society in social science. Russian civil society, as measured by group membership, hours of voluntarism, money donated, petitions signed, protests attended, and so forth, looks anemic by comparison. But this theoretical construct of civil society is rife with contradictions. First of all, are we really prepared to say that the Soviet Union had *no* civil society, just because it did not resemble the West? And what do we do with those parts of Western civil society that do not themselves conform to this ideal type: those undemocratic and exclusionary activist groups? These concerns and others have led some scholars to dismiss the very concept of civil society as too unwieldy a term to be of use (Viterna, Clough, and Clarke 2015). Rather than dismiss the study of civil society as a failed endeavor, perhaps it is only that the ideal type has misinformed us, and we must rethink and re-examine an old construct in a new light.

The time has come for a new theory of civil society, one that is developed inductively through a close case study of a rich, complex, and varied context. Using ethnographic methods, both historically grounded and richly descriptive, I examine the endeavors of local environmental groups to protect and preserve Lake Baikal in Siberia. In doing so, I re-conceptualize the civil society construct. Specifically, I view civil society as participating in *a field of power* along with political and economic actors. The strength and form of civil society is always relational to other players in the field of power; civil society formation is mutable and historically contingent. Although abstract, the field of power becomes manifest in the various struggles and interactions of local civil society groups in their attempts to create social change.

Environmental activism in Baikal has traversed rapidly changing social, political, and economic contexts. It arose under the authoritarian government of the Soviet Union, it survived the state and economic collapse in the 1990s, it experienced and exploited the opened borders of globalization, and it has navigated the resurgent authoritarianism of Vladimir Putin. By delving into the depths of Baikal environmentalism, through this turbulent history, and in this special and unique location, we can pose new problems and construct new solutions to age-old questions about civil society.

The Civil Society Concept

"Civil society" is a loaded term. For the Kremlin, civil society is a harbinger of chaos with revolutionary intent. It is a force that ends governments, and it is well worthy of their wary distrust.[3] The West, on the other hand, views civil society as so unqualified a public good that governments allocate state funds for its promotion. An entire sector of Western national economies—the nonprofit sector—is devoted to voluntary, donation-based service-centered efforts.[4] The United States even considers civil society to be part of its national DNA, the life-blood coursing through a "nation of joiners" (Tocqueville 1981).

In the scholarly community, civil society has a similarly complicated story. While much has been written about it, especially over the last three decades, there is little consensus over what exactly civil society is or what it does. For some, it is the very foundation of democratic governance (Putnam 2000); for others it is the exclusionary underbelly of pluralist modernity (Alexander 2013). While it is a concept that seems to demand scholarly attention, it also evades easy explanation. Its contribution is indeterminate and its nature ineffable.

Yet, for all the ambiguity surrounding civil society, there exists a "broad brushstroke" rubric, which guides our understanding and study. Civil society comprises all the shared human activities that take place outside of the government, the economy, the individual, or the family. It represents the outpouring of voluntary human initiative, wherein people unite and act in concert toward some self-directed end.

Political theory ascribes to civil society a special status as the guarantor of democracy. Free association is seen as the cultural expression of a democratic ethos. Tocqueville (1981) spoke of civil society as fostering "habits of the heart," that in turn encourage collective responsibilities and affective ties among the citizenry. For Almond and Verba (1963) it formed a "civil culture" that carried over into political life. Putnam (1994, 2000) similarly showed a positive relationship between social capital and various pro-democratic outcomes. Skocpol (2003) studied the history of American social organizations, finding that the broad base of their membership had important democratizing potential.

Civil society was also the term used by dissidents in Eastern Europe and Latin America to describe themselves and the public lives that they yearned for in their home countries. Vaclav Havel (1978) made an impassioned appeal to the need for "the independent, spiritual, social and political life of society," and the belief that government should rise up from the grounded experience of human life rather than impose ideological and social structures from above.

Gellner (1994) likewise described free functioning civil society as the very "condition of liberty" in Eastern Europe and in the world. Following the anti-authoritarian revolts erupting across Eastern Europe and Latin America in the 1980s, scholars came to see civil society as a force to curb the concentrated power of the state (e.g., Arato 1981; Diamond 1994; Kim 2000).

As globalization progressed in the new millennium, scholarly attention also turned to the ways that civil society has worked to check the power of corporations. Social movements have succeeded in forcing the enactment of global standard practices in the environment and labor rights (e.g., Bartley 2003, 2007). Social organizations fill the gaps that corporate dominance and neoliberal governance leave in their wake (Steurer 2013; McKeon 2015). Civil society works to keep corporations accountable and to demand adherence to principles of justice.

However, despite the widespread scholarly acceptance that civil society plays a democratizing role, there is a troublesome contradiction between this claim and the empirical reality of a civil society that is itself undemocratic. Lynch mobs, pogroms, populist movements, and so forth comprise the so-called dark side of civil society (Alexander 2013). Early democratic theorists were as frightened of the ability of willful majorities to trample the rights of minorities as they were of the despotism that preceded democracy (Mill 1978; Madison 1987). Voluntary association in, for example, the Ku Klux Klan is a rather doubtful guarantor of democratic governance (Kaufman 2003). Nationalism and religious fundamentalism take place within civil society, but their contribution to the creation of the "good society" is highly questionable.

Neither is there agreement on the means by which civil society is constituted. For some thinkers, it is a phenomenon that arises through cultural interaction. It is a discursively defined sphere with varying degrees of participation and inclusion (Habermas 1985; Kaldor 2003; Alexander 2006). However, if civil society were merely discursive, then why would dissidents under repressive regimes claim it was a liberty denied to them (Havel 1978)? Civil society was posited as something lacking in these authoritarian regimes, something without which tyranny could flourish (Gellner 1989). As an institution, civil society requires more than a discursive origin. Therefore, Cohen and Arato (1993) determine that civil society flows from legal rights and freedoms, such as speech and association. Unfortunately, this conclusion yields a strange tautology, whereby the "condition of liberty" is itself derivative of the liberal state. Finally, neither the cultural nor the legal explanation of civil society adequately accounts for the phenomenon of global civil society and the transnational interactions that take place beyond sovereign states and cultural boundaries (e.g., Walzer 1995; Florini 2000; Colas 2002; Clark 2003; Kaldor

2003; Laxer and Halperin 2003; Friedman, Hochstetler, and Clark 2005; Eberly 2008, Smith 2008).

In this book, I argue that most understandings of civil society are fundamentally flawed. Too often, civil society is set apart conceptually from the other social actors that it engages. Most scholars describe civil society as "a check" on concentrated elite power, either in the form of an overbearing state or a self-interested corporation. Essentially, when contemporary scholars discuss civil society and its role, by repeatedly referring to it as a "check" on elite power, they pay an unacknowledged debt to Montesquieu (1989), the eighteenth-century political theorist who first conceived the system of governmental checks and balances.

These Montesquieuian theories of civil society are not wholly inaccurate—but neither are they complete. "The separation of powers" is a criterion in the *constitution* of a state; governments can be designed and established in accordance with the principle of checks and balances. However, civil society is not created according to a predesigned constitution. Legally enshrined freedoms, such as the freedom of assembly or of speech, can shape the form that civil society may take and can grant it greater or lesser freedom to act, but civil society exists *sui generis*.

Civil society is not simply a "check" on the power of economic and political elites; it is a power unto itself. Participants in civil society are just as eager to put in place their own vision for society as they are to check those of state and corporate opponents. But a failure to check these alternate powers does not inherently imply consent. Neither does it mean defeat. Instead, actors in civil society are engaged in a dynamic, ongoing struggle to determine the course of societal development. For this reason, civil society is better understood as one player among many, where each is endowed with different resources and all operate from their unique social positions. To analyze civil society in such a manner, we must turn to field theory.

Field Theory

Social scientists, long accustomed to examining discrete and somewhat disembodied "variables" to achieve social explanation, have increasingly come to embrace a new analytic method that places individuals and groups within an interactive field. A "field" is a place-metaphor for a space of social interaction. Scholars have been slowly building upon and reconceptualizing the metaphor since the late 1960s, and while there are differences between these many definitions, a general consensus exists on the usefulness and applicability of the construct at this most basic level.

A field is a space of social activity where actors compete or cooperate to achieve their own desired ends relative to the general thematic project of the field—the "stakes" of the field. For Bourdieu (1969, 89), fields comprise "systems of relations between themes and problems" wherein agents endeavor to take positions relative to one another. Turner (1974, 135) views the field as "an ensemble of relationships between actors antagonistically oriented to the same prizes or values." Martin (2003, 20) refers to them as "fields of organized striving," while for Fligstein and McAdam (2012) they are fields of "strategic action." The field is defined by the shared understanding of those involved as to 1) the existence of a relationship and 2) the rationale for the relationship. Usually, this rationale is described as a shared understanding over what is at stake and the general (albeit changing) rules that govern social action in the field.

Although Martin (2003) harkens back to gravitational and electromagnetic fields to explain field theory in the social sciences, I would suggest that a superior "natural-scientific" analogy can be found in ecology, the study of ecosystems. Field theory emphasizes relationships. No piece of the system—its actions and motivations—can be fully grasped without accounting for myriad other interests that all seek to survive and thrive in a given place. Here, the topographic imagery of a "field" becomes especially apropos. While in the natural world, ecosystems assume some geographic proximity, in social systems, the field is "visible" through interpersonal relationships and a shared imaginary that may even span the globe.

The theorist most closely associated with social fields is Pierre Bourdieu, and he has elaborated his analytic method over a number of seminal works in a variety of contexts (e.g., Bourdieu 1983, 1984, 1990, 1993; Bourdieu and Wacquant 1992). Bourdieu likens social fields to sports fields, and individuals in the fields can be considered the players. Another metaphor he provides is that of a battlefield where the actors are combatants. In both analogies, players and combatants occupy different positions on the field that define their perspectives and their opportunities. So, too, with social fields. Whether actors are in the field of fine art, literature, academia, finance, or so forth, they jockey for advantage—either to change their own position or to elevate the status of their present position relative to that of others. To do so, they utilize whatever resources are at their disposal. Bourdieu refers to these resources as different types of capital: economic, social, cultural, symbolic, and so forth. These "species of capital" are field specific and may be carried from one field to another, although the exchange rate varies considerably. For example, a rich Rolodex in the world of Hungarian folk dance commands status in that particular field but garners little value in the field of, say, ornamental horticulture.

Bourdieu's studies have mostly focused upon individuals. His fields are schemata of stratification. The very notion of "capital" as a resource suggest that it yields "returns" in kind—providing to its possessor augmented amounts of the same. But field activity is not solely geared toward status attainment, and people are not always striving toward individualistic goals. Collectivities, groups, and organizations likewise operate as actors in social fields. Organizational theorists, like DiMaggio and Powell (1983) and Meyer and Scott (1992), were pioneers in describing organizational fields that operated according to an underlying "institutional logic"—a shared understanding of the field and relational behavior among corporate actors. Go (2008) showed how entire countries changed their imperial behavior in response to global field logics. Fligstein and McAdam (2012) made group behavior and collectively defined goals a major component of their general theory of "strategic action" fields. Voluntary organizations and nonprofit nongovernmental organizations (NGOs) also engage in "organized striving" (Martin 2003) toward a particular social vision that they have collectively defined. Field theory, therefore, can be usefully applied to the activities of civil society.

However, according to field theory as it has been previously elaborated, there is no discernible "field" that would encompass the entirety of civil society. Instead, individual organizations are seen as interacting with other organized social entities in issue-specific fields. By this logic, an environmental organization would not occupy the same field as an HIV/AIDS organization, but each would share their field with actors outside civil society. Field theory may only be employed to explain whether and how these "atoms" that make up the abstract collective that we refer to as "civil society" achieve their own unique ends in their own specific fields. About civil society as a whole, it generally tells us nothing. I argue that, on the contrary, field theory can tell us about civil society as a whole, but in order to do so, we must look beyond the intentional and "strategic" actions of particular organizations in their issue-specific fields. This definition of what a civil society group "does" is simply too narrow. There is a broader way to conceive of civil society and what it does in the world. There is a reason that scholars have latched on to the general concept of "civil society," creating a single term to lump these disparate and distinct groups into one large pool. For all their differences, these groups and organizations do have something in common. They share something that sets them apart from atomic individuals in their private lives, something that renders them more akin to formal, macro-level institutions, such as the state or the economy. Fundamentally, what these civil society groups share is power and the ability to wield it. When we see civil society as a wielder of power, it makes manifest a field of social activity that I call *the field of power.*[5] The field of power presents a new lens by which to examine civil society, one that is better prepared to meet the challenges of the twenty-first century: an age of globalization and resurgent authoritarianism.

The Field of Power

The field of power is a meta-field. In the field of power, the stakes of activity are not confined to any particular outcome in any particular field. Rather, the field of power is the space where meta-powers confront one another. At stake is the ability to act toward achieving one's end—to put in place one's social vision—regardless of the social field at hand. Essentially, what is at stake is power itself. Like any social field, it is a space of dynamic interaction: the field is historically situated, globally embedded, and highly contingent. And because it constitutes a conflict over power itself, the field of power is manifest in any social field and may impact every social field, to a greater or lesser degree. In this book, I carefully examine one particular field—the field of environmental protection at Lake Baikal—to examine the larger field of power and trace its activity. By tracing the interactions of various parties, each with competing visions for Baikal, across time and through space, we may behold the contours of this power field. Civil society is but one player, albeit an important one, in this field, and it is continuously shaped in relation to other power players in historically situated moments, charged with contingency.

Who may contend within the field of power? In order to play on the meta-field, one must wield a meta-power. A meta-power is a power source that can exert influence in virtually any social field. It is a power whose influence is fairly constant across all fields. These are generalizable power sources. In this book, I identify three such meta-powers at play, and they are the respective arsenals of the three principle players in the field of power: the state, the corporate elite, and civil society.[6]

The first two should come as no great surprise. Classical sociological theory has attuned us to these generalizable powers. Max Weber defined the state as the entity that alone holds the legitimate use of force (Gerth and Mills 1958) and, with it, the ability to determine the law. The state defines what is permissible and impermissible in a given polity. Thus, legal power acts upon the entire social body, up to the geographic limit of state sovereignty. The law is a generalizable power. For the economic elite, their meta-power is financial. Karl Marx (1964, 1976) devoted much ink toward the recognition that money is an abstraction of exchange-value. Money can and does play a role throughout all sectors of society as a general means of trade. It may be traded for virtually anything in all social fields. Money, held in large and ever-renewing concentrations by capitalist elites, is also a generalizable power.

What about civil society? It too has a generalizable power source, which renders it a formidable opponent to political and economic elites in the field of power. The power of civil society can be found in *human beings*. Just as money and the law can be found in all social fields, human beings similarly populate social fields, and their ability to act voluntarily and in concert gives civil society a unique and generalizable power source.[7] Often, this power is reduced to mere capacity—"skilled actors" (Fligstein 2001) in social movements and nonprofit organizations who possess, in varying degrees, the knowledge and infrastructure to mobilize the public (McCarthy and Zald 1977): they sway public opinion, get bodies on the streets in protest, produce signatures for petitions, or foster boycotts and divestment campaigns.

But capacity alone is not the source of civil society's power. Still more importantly, civil society has power that comes to it by virtue of its *virtue*. To say that civil society garners power from virtue is not to suggest that civil society actors are virtuous people, that their goals are good, or that the outcomes of their actions are necessarily beneficial. Rather, it is to explain what these groups have been *socially deemed* to possess that differentiates them from their opponents in the field of power. People build civil society voluntarily; they contribute their efforts of their own free will, and they do so because they have deemed the cause to be a "good" one. Civil society holds, in the public mind, a kind of "worthiness" (Tilly 2004). There is a form of power that accrues to civil society through its widespread voluntary social engagement. The wider the reach of engagement, the greater the "virtuous" civil power. In an era where "right" and "wrong" are neither explicitly defined nor universally shared, when our sense of good and ill is untethered and shifting, civil society can show us what society values. Civil society lays bare those causes people value so highly that they will give, as unencumbered offerings, their time and treasure. Civil society is a morality for modernity.

Once we understand how worthiness functions for the creation of civil power, we no longer need to tie ourselves into theoretical knots to explain civil society as both the heroic guarantor of democracy or the villainous purveyor of social exclusion. There is no Manichean divide between a "Good Side" and "Dark Side" of civil society. Rather, the wide plurality of civil society groupings embodies the range of socially defined moral virtue at any given time. The more broadly shared the cause, the greater the civil power it can command. The greater the worthiness, the more generalizable the power.

Thus, civil society's power is the ability to move human beings, through capacity and through virtue. Because there are human beings throughout society, this power is also generalizable. As such, it operates as a potential threat to those who seek to dominate by means of force or wealth. The latter two can similarly mobilize resources to enact their respective visions, but neither have civil society's claim on virtue. That quality is a power that the state, as an allegedly neutral arbiter, and corporations, which operate to maximize personal profit, do not and cannot inherently possess.

Understanding the field of power and the meta-powers at play helps to explain what has hitherto remained a general assumption: that civil society is "a check" on political and economic elites. Theoretical outlines of civil society reinforce this assumption. In their comprehensive theory of civil society, Cohen and Arato (1993) also build a triadic model that places civil society counterpoint to the economy and the state. Wright (2010) offers up a similar schematic. In part, this assumption is based upon empirics: the vast majority of the literature on civil society names either economic or political elites (or both) as the chief targets of civil society organizations. However, until now, there has been no clearly enunciated theoretical rationale for why this is so. The power of the state and the corporation are simply acknowledged as self-evident, and civil society's antagonism to their concentrated power needs no further justification. But when we analyze them together as players in a field of power, the reason that *these particular players* are the most central becomes clear. It is not the players that are important, but rather the arsenal at their disposal.

Once we understand the field of power as the space to deploy generalizable power sources, we can also see how players use their own arsenals to commandeer or disable the arsenals of others. I refer to these as "plays" in the field of power. While many of these activities take place within smaller, more specific social fields, they are plays in the meta-field when the action serves to alter the balance of power among these three generalizable power sources. When actors in civil society advocate for new laws, they are using civil power to lay claim to legal power. When they campaign for divestment or organize a boycott, they are deploying civil power to disable an economic arsenal. Because civil society is a power in itself (as well as a collection of smaller, more specific fields), the other players may use their respective arsenals to similarly acquire or curb civil power. They may "colonize" civil society, using civil power for "uncivil" purposes, or they may suppress civil society by constraining its scope. They may also target its virtue, in an attempt to cripple its unique claim to speaking for the public good. These actions may take place in particular fields, but they are plays in the field of power.

Viewing civil society as a player in a field of power has a number of advantages over existing models of civil society that are more static in structure: 1) it considers civil society to be a source of power in its own right, rather than a mere check on other powers; 2) it allows us to look at the strength and weakness of this power source relative to other power sources; 3) it enables us to trace changes in these relative and relational powers historically over time; and 4) it brings multiple spatial scales into play in the field of power. These advantages make the field of power approach particularly applicable to the challenges of social analysis in the twenty-first century. We are now firmly entrenched in the age of globalization, but we have witnessed in its forward march a renewed ambivalence toward democratic governance.

The Field of Power in Context: Globalization and Authoritarianism

During the Cold War, the word "totalitarian" could be heard echoing through the halls of academe, bandied about by policymakers, and repeated by the media to describe the geopolitical and socialist-economic phenomenon that was the Union of Soviet Socialist Republics (USSR).[8] The word conjured up a monolithic monster: a uniform centralized state that controlled every aspect of people's lives, having stripped the flourishing individuality from its "masses" through a singular, all-encompassing ideology and the terror of a violently repressive disciplinary apparatus (Arendt 1966).

Once the Cold War ended, and the ideological clouds dispersed from analysts' eyes, it was possible to discern that totalitarian theory described a fiction. Even the strictest and most repressive dictators cannot wholly dominate human agency. Quite the contrary: Soviet citizens created their own worlds, in accordance with (or in antithesis to) the centralized state. They sought to make their own meanings and understandings of official ideology in their personal lives (Hellbeck 2006) and they fashioned personal, as well as political, forms of resistance (Kenney 2002; Yurchak 2006). Interestingly, environmental protection was among the first of these "little corners of freedom" to be rediscovered by Western scholarship (Weiner 1999). Indeed, it became apparent that socialist states had their own conservation ethic (Brain 2011), one that engaged the citizenry in pro-environmental behaviors (Gille 2007). Rather than the anomic masses of Arendt's (1951) theorizing, Soviet society was a mess of collectivities that were engaged in their own reflexive and self-directed actions, including environmental ones.

However, it is one thing to acknowledge the existence of a civil society in the Soviet Union, and another to imagine away the very real constraints that it faced. If the Soviet state failed to orchestrate wholly the lives and behaviors of its citizens, it was not for lack of trying. From propaganda and censorship, to rationing and five-year plans, from internal passports to secret police, the imagined space for social action in the Soviet Union was subject to the pull and push of exogenous forces even while activists were pursuing intentional aims. In this way, Soviet civil society is not unique, but the field of power that shaped it was, which allows us to consider the different formations of civil society across time and space.

The field of power pushes our analysis of state-society relations beyond that of mere "resistance"; it also prevents us from falling into the totalitarian trap, overemphasizing the dictates of the dictator. Because the state writes the rules for social activity within its sovereign territory, it can write itself into any social field. Also, the state has the only legitimate use of force to enact the rules of its own creation. Because of the obviousness of this form of power, there is a tendency to reduce the field of power to the mere struggle for state control. Bourdieu (and Wacquant 1992, 114) and Fligstein and McAdam (2012, 67) make this same mistake. The field of power is emphatically *not* equivalent to the political field or any subfield thereof. If this were the case, then totalitarian theory would, in fact, win the day. The law is a meta-power, but we must not fail to see the other meta-powers at play and their evolving interaction. The field of power can be dominated, but a *field* it decidedly remains. As Russia retreats once again away from democracy, there is a renewed public and scholarly interest in Vladimir Putin as the figurehead and the absolute authority across his sovereign territory (e.g., Sakwa 2007; Shevtsova 2007; Gessen 2012; Arutunyan 2014; Sperling 2014; Tsygankov 2014; Dawisha 2015; Gelman 2015; Laqueur 2015; Myers 2015; Garrels 2016; and Zygar 2016). But the action and activity in the Russian state is more dynamic than the will of the dictator, and the story of Russia is not simply that of Putinism.

Moreover, the field of power does not stop at Russia's borders, even if Putin's sovereignty does. The importance of globalization in the post-Soviet era cannot be overstated. We are interconnected in flows of goods, ideas, and people as never before in human history. Even the most local of fields can utilize multiple spatial scales to enact power plays. Putin may be Russia's dictator, but he is a twenty-first-century dictator who must contend with the promises and perils that globalization brings about.

Finally, there is an important lesson that the field of power offers to the study of civil society in general, and that is the recognition that *all* civil society is relationally constituted within a field of power. Too often, civil society in the post-Soviet region is bemoaned as a distorted version of the "ideal typical" civil society, which is often enshrined by the Western NGO sector. However, a field of power approach can dismiss any such "ideal type" as folly. Western civil society, although legally freer than its counterparts in other locales, still confronts a web of national and global governance institutions, as well as the ever-widening reach of corporate capitalist globalization. It is no less shaped by its own field context than are its brethren in autocratic regimes. The dynamic power relationships of Western civil society may be less visible, given the more familiar—one might even say hegemonic—context, but the field of power is present nonetheless. Therefore, civil society is not a construct of legal rights and freedoms, as Cohen and Arato (1993) maintain; it is a people-power that necessarily confronts other meta-powers in an overarching, and highly contentious, field. The field of power allows for analysis of this complexity wherever it is found.

The field of power need not be globalized, but ours is a global age. Globalization adds a new dimension to older conceptions of civil society, as well as its relationship to authoritarian states. In the chapters that follow, I show that even local groups contending with place-based problems, such as the protection of Lake Baikal, are necessarily bound up in national and transnational phenomena, and the intersections of these spatial scales impact the interplay of arsenals in the field of power.

This book examines the interplay of these powers in relation to local environmental civil society. It looks at globalization in Siberia, a land that stands as a metonym for isolation and exile. It confronts authoritarian politics in a country that has been buffeted by ideological extremes. It follows chains of ownership and ideology that cross continents, and it traces the footsteps of individuals along similar transnational trajectories. Reaching from San Francisco, South Lake Tahoe, Seattle, London, Berlin, and Moscow, these pathways finally intersect and touch down in the middle of Siberia, in the city of Irkutsk, beside Lake Baikal, the Sacred Sea.

A Case Study of Environmentalism at Lake Baikal

Baikal is a unique ecosystem, home to thousands of rare endemic species. It is a scientific and aesthetic wonderland, and as such, the lake has become a focal point for environmentalist attention in Russia and around the world. Activism to protect Lake Baikal stretches back into the Soviet era, and it continues through the present day. Much of this activism is based in the city of Irkutsk, the regional capital, located forty-five miles from the shore of Baikal. Irkutsk is no stranger to citizen engagement and dissent. As a distant outpost of the Russian Empire, the city became home to countless political prisoners and exiles, including the famous Decembrist revolutionaries of 1825. Irkutsk is also a center for scholarship and higher learning, hosting a substantial number of universities and research institutes. This potent combination of political dissidence and scientific inquiry combined in Irkutsk in the mid-twentieth century to foster the first mass environmental movement in Russia, with the aim of protecting the unique lake, Baikal. The movement has continued to exist in one form or another through the decades, stretching across the Soviet/post-Soviet divide. Because of the importance of Lake Baikal, the region has attracted myriad international actors who collaborate with local activists in Irkutsk, rendering the area a permanent, if volatile, transnational activist space. Thus, Baikal is a site that captures the historical, political, and transnational dimensions of local environmental civil society in Russia. Through Baikal, we can find in concrete form the interplay of power holders, as each seeks to enact its own vision for the lake and its environs.

There are many types of civil society in Russia, each of which confronts the question of globalization and authoritarianism, and the content of each group necessarily moderates the encounter. My focus is on environmental organizations, whose subject matter brings important advantages. First, the environment is, by nature, borderless. Climate change is a global problem. Regardless of its source, carbon dioxide mixes freely in the planetary atmosphere and its effects are felt worldwide. Even end-of-pipe pollution seldom stays comfortably near the sullying enterprise, as problems like acid rain have shown. Use of one of the few remaining global commons—the open sea—has witnessed dangerously low levels of fish stock and a floating island of degraded plastic in the Pacific Ocean that is larger than the state of Texas. The political realities of transnational pollution have resulted in an increasing number of treaties and international cooperative agreements aimed at addressing the disconnect between cause and effect when one country's practices impact another country's ecosystem (e.g., The Helsinki Convention on the Protection of the Marine Environment of the Baltic Sea Area; the Kyoto Protocol; the Great Lakes Compact; see also Darst 2001). For a study of civil society's engagement within a globalized field of power, there is no cause as truly transnational as planet Earth.

The other reason to hone in on environmental organizations is the world-historical importance of their subject matter. Environmental issues are the defining issues of our time. Since the dawn of industrialization, the ability of humans to impact the natural world has increased dramatically, and in the twenty-first century the legacy of this damage has reached crisis levels on multiple fronts. Global climate change as a result of burning fossil fuels is currently set to raise temperatures an average of 4.5 degrees Celsius by the end of the century—more than double the number of degrees of warming that is considered ecologically safe by climate scientists. Industrial pollution and toxic chemicals remain untested and under-regulated. Deforestation and desertification continue at alarming rates, impacting carbon sinks, biodiversity levels, and access to fresh drinking water. Meanwhile, the world is facing the greatest mass extinction since the dinosaurs, due largely to anthropogenic habitat loss. Human civilization cannot continue at its present pace, using present means, without undermining its material existence and posing an existential threat to itself and the natural world that surrounds and supports it. There is no social realm that is as in need of a flourishing civil society as this one.

The source of the contemporary environmental crisis is the production of goods for the fulfillment of human needs and wants. However, the solution to environmental crisis is not likewise simply a matter of the mode of production. The wanton environmental destruction perpetrated by the Soviet Union should put to rest any suggestion that ecocide is solely a property of capitalism. Likewise, we need not seek explanations for Soviet environmental degradation strictly within its own unique productive system. In their exhaustive review of Soviet environmental crimes, Feshbach and Friendly (1992) list a relentless stream of crises in land, water, air, and—subsequently—human health. In many instances, they and their native informants claim that the problem stems from the command economy. Without market prices, they explain, there is no way to assess the value of squandered natural resources.

What Feshbach and Friendly (1992) conveniently forget is that the capitalist West had functioning market prices throughout the nineteenth and twentieth century and still managed to degrade its environment and bring forth its own ecological catastrophes. The market did not save the West from the errors of the Soviet Union. If nature was less ravaged in the United States and Western Europe, it was not due to the market but rather to the collected efforts of committed individuals, determined to prevent the wanton abuse of nature and its gifts. It was civil society's intervention in the normal "treadmill of production" (Schnaiberg 1980) that ebbed the speed and scale of environmental destruction in the capitalist West, not market mechanisms. Nearly every progressive environmental action taken in the United States and Europe came at the behest of environmental activism, not the magic of the market. When we contrast the state of the environment in the Soviet Union with that of the "Free World," the principle variable at play is not production, planned or otherwise. It is the relative powerlessness of Soviet civil society that set it apart, environmentally, from the West. Any attempt to understand environmental degradation must begin with a discussion of power.

Data and Methods

To chart the terrain of the field of power, I examine local environmental civil society in globalized, yet increasingly authoritarian, contemporary Russia. I spent ten months conducting field work in Irkutsk, a regional capital city in eastern Siberia. My primary method was ethnographic, but it is an ethnography that takes seriously the historical processes in which human activity is embedded (Comaroff and Comaroff 1992). At the same time, I draw upon Michael Burawoy's ethnographic methodology, which pays attention to broader social structures operating on multiple spatial scales, all of which act upon and become visible in the local context (Burawoy 2000; Gille and Riain 2002). My analysis intertwines local observations, national and transnational intersections, and the historical moment in which events are taking place.

During these ten months in Irkutsk, I immersed myself in the multi-organizational field of environmental activism in the region. Throughout the book, I refer to my informants as "activists," although I suspect that some would eschew the label. They may not all carry placards in the streets, but these individuals participate in projects with an agenda: namely, to foster a flourishing and sustainable ecosystem around Lake Baikal. They are members of the region's environmental civil society, which I define as those social groups and networks that cooperate and coordinate activity designed to promote environmental sustainability through a wide variety of means.

I focus particularly on the three strongest organizations in the community, and these also have the most transnational connectivity: the Great Baikal Trail, the Tahoe-Baikal Institute, and Baikal Environmental Wave. The bulk of my data come from participant observation. In addition to fieldwork in Irkutsk, I traveled to partner organizations in Moscow, Seattle, San Francisco, and South Lake Tahoe, following the footsteps of people and the flow of resources from these distant sites to the shore of Lake Baikal. I supplemented these observational data with fifty-two interviews with key stakeholders, including environmentalists, government workers in the system of protected territories, corporate sponsors, and transnational supporters. Archival research illustrated the contours of late Soviet environmentalism and the activities of environmentalists in the transition period. The data collectively tell a story of civil society invigorated and imperiled as activists find themselves pulled by the strong tides of a volatile political economic structure: one that is increasingly authoritarian nationally but unrelentingly global in scope.

About the Chapters

I begin in Chapter 2 by setting the geographic stage for the study and introducing readers to Lake Baikal. I trace the history of various environmentalist activities around the lake relative to the field of power. Civil society in the Soviet years can be characterized as existing in two threads: the formal and the informal. On the one hand, there was the Communist Party-sanctioned All-Soviet Society for Nature Protection (VOOP) and the Komsomol student nature protection movement, the *druzhiny*. But around Lake Baikal, there was also an informal and spontaneous eruption of activism that persistently fought the state throughout the construction and operation of the notorious Baikalsk Pulp and Paper Mill. The field of power was dominated by the Soviet state and the military-industrial complex, with a constrained space of activity for environmental civil society. The environmental outcomes of concentrated power within the field were dramatic.

When the Soviet Union collapsed, local environmentalism made a radical shift in form and structure with the advent of independent association and international collaboration. Chapter 3 discusses the development of Baikal activism relative to the field of power, across spatial scales and over time. It follows the founding and development of the three principal environmental NGOs in Irkutsk: Baikal Environmental Wave, the Tahoe-Baikal Institute, and the Great Baikal Trail. Each was created by means of transnational connections and supported by foreign funding. They were also founded at different points in Russia's transitional trajectory—two emerging out of the Soviet period, and the last founded in the Putin era—which also influenced their form and focus.

Chapter 4 gets to the heart of the importance of globalization for civil society. Following one particular project—a webinar series between villagers around Lake Baikal and residents near Lake Tahoe—I show the tremendous impact that cross-cultural connectivity has on the perceived horizons of social possibility. Domestic activists cannot spur villagers to belief in their own potential to produce change, but over the course of the webinars, exposed to a different culture and context, they begin to believe in their own abilities. Even if the intended effect of the exchange fails, the very act of communicating across cultures has a profound influence on social imagination, which is an important precursor to social change.

Chapters 5 and 6 introduce the multinational capitalist corporation into the field of power. Chapter 5 discusses the transition years and the rise of the Russian oligarchs through the person of Oleg Deripaska, among Russia's richest men and a leader in finance and industry. Deripaska controls the En+ Group, a holding company for a variety of extractive and industrial enterprises. As he and his peers move into the global marketplace, they learn to conform to global business practices, including corporate social responsibility and cause-marketing. En+ becomes the chief corporate sponsor for environmental activism in the Baikal region. I discuss environmentalists' ambivalent reactions and the equally ambivalent environmental outcomes.

Chapter 6 focuses upon particular projects enacted by En+ in the Baikal region, often with the cooperation and collaboration of local environmentalist groups. Each of these projects is an ideological gambit, but they can be weighed one against the other for the degrees of freedom they allow. Corporate entities, such as the En+ Group, embark on value socialization projects that seek to constrain civil society so that it will not threaten business interests.

Chapter 7 brings the state back into the mix in the field of power. Globalization is a threat to sovereignty, and authoritarian regimes are deeply dubious of the influence of foreign forces on domestic civil actors. In Russia, following a massive protest against Putin in 2011, the state passed the "Foreign Agent" law. Any nonprofit organization that receives funding from abroad and engages in political activity—broadly defined—is deemed a Foreign Agent. I trace the fallout of this law in the field of environmentalist groups in Irkutsk, showing how the law targets particularly threatening organizations, to establish a restricted, non-threatening civil sphere.

The book's conclusion in Chapter 8 addresses the multi-layered and historically contingent field of power around Lake Baikal in Russia. As Putin strives once again to dominate the field of power, as did his Soviet predecessors, the ambiguities inherent in the era of globalization become more apparent. Moreover, our exploration of the field of power around Baikal can offer important lessons to the West, which must face different dominating forces. Finally, the field of power, while an abstraction, has material consequences. Even if the field is dominated, the world that it shapes is changing. These changes ensure that new dynamism will eventually be brought to the field, and with it will come new opportunities for environmental civil society to exert its influence—if it is not by then too late.

2

Lake Baikal

While visiting Ulan-Ude, the capital of Buryatia, to interview environmental activists on the lake's eastern shore, I stopped into a *poznaya*[1] for dinner. The little café was about fifteen feet by ten feet and had only eight tables that were flimsy contraptions. I went up to a window in the wall to place my order: two *pozi* and a cup of black tea. The lady behind the window informed me that it would be a twenty-minute wait for the *pozi*, but I assured her I was in no hurry. I chose the smallest of the tables against the wall and began scribbling notes from my recent conversation.

The *poznaya* was not empty. There was a party of five men, ranging in age from early twenties to late forties, two Russians and three Buryats. It was evident that they had been drinking and were still at it. Their jocularity and frivolity were also entwined with bouts of anger, and the volume of their ejaculations was consistently increasing. A *babushka*[2] cast hateful glances in their direction each time their noise level increased. A mother with two small children finished her meal and scurried her brood past the raucous table, scolding the men for their language as she passed. I was also beginning to wish that I had not agreed to wait the twenty minutes for the *pozi*.

One of the men at the table began to call out in my direction, "*Devushka! Devushka! Ty menya ni znaesh, chto li?*" [Girl! Girl! Don't you know me?] I buried my face in my field notes and pretended not to hear. The calling continued, and I continued to ignore it. Suddenly, the voice was standing beside me—a Buryat man with a large build, in his late thirties. He swayed slightly and spoke too loudly for comfort, his drunkenness showing through. "Why didn't you call me?" he demanded.

"Leave me alone," I replied.

"Don't you remember me? You promised you would call and you never called."

"I don't know you," I said. "We are not acquainted."

"Yes, we are. We met at the discothèque in November. Don't you remember me?"

"Here in Ulan-Ude?" I asked him.

"Yes," he said.

"That is impossible," I told him. "Today is the first time I have ever been in Ulan-Ude. I have never been here before."

"Really? Today is your first time in Ulan-Ude?" I nodded, and began to worry I had said too much. "Forgive me. You look just like a lady I met in November. So what brings you to Buryatia?"

I told him that I was studying environmental protection around Lake Baikal (Fig. 2.1). His eyes suddenly widened.

"Baikal? You are here to study Baikal?"

"Yes."

"Let me tell you about Baikal!" He said, now seating himself beside me, his tone suddenly very serious. "People come from all over to study Baikal, but they will never know it. It is immeasurable! They try to measure it, but they can't. It is always changing. Do you know how deep it is? You can't even imagine it, it's so deep!" His slurred speech punctuated certain words with enthusiasm and wonder. "Baikal is incomprehensible! So you should just go home. You came here to study Baikal, but you will never know Baikal. Never."

Figure 2.1 Lake Baikal in the summer. Photo Credit: Author

Baikal occupies a unique place in the Russian psyche. Everyone knows Baikal, and increasing numbers of Russians make the pilgrimage to see it. It is a landmark of national pride (Rasputin 1996). The name "Baikal" is affixed to various Russian products, from computer chips, to vodka, to firearms. For those who live near its shores, the lake is regarded with reverent awe, bordering on religious mysticism. The lake is frequently personified and treated as a kind of god or supernatural entity (Mathiessen 1992). Themes about the lake's unknowability and its life force weave through discourse about Baikal. In addition to the ecstatic drunk quoted above, a number of other conversations about Baikal that I observed during my time there reiterated the very special attitude locals hold for this treasured place. The following are two more such examples I recorded, from a teacher in Irkutsk and a librarian in Baikalsk:

> TEACHER: For me, Baikal is that place where all life's worries and internal stresses [disappear]. Thanks to Baikal, I can quickly calm them. So, for me, first and foremost is the living essence that is Baikal. For me, Baikal is alive. Not only that, he is my friend, with whom I come to talk, and who will understand, who will support me and give me strength and energy. Baikal is many things for me. Most importantly, when I travel places, when I am abroad, I am always connected to it. I always want to return to it. Because, when you spend time at Baikal, you will never see the same thing. It is always different. It is, for me, something incomprehensible. It is impossible to ever really know it, to the end, because it is always changing. This is his character.

> LIBRARIAN: The idea [behind shamanism] is that there are different spirits that reside in nature and in particular places. On Baikal, this is particularly obvious and visible [yarko i vidno]. Baikal is a living thing, a living organism. He thinks, he understands. The Buryats and the Evenks understood this. No matter how much we study the Lake, we can only understand 1/1000th of it. Baikal is patient and loving, but he also has power. And he has great patience with us for our stupid acts. So far, Baikal has been patient with us, but maybe not forever. He hasn't punished us yet, but he is great and he could punish us. But so far he is patient.

If Baikal looms large in the public mind, there is good reason. Lake Baikal is like no place else on Earth.[3] The lake and its environs are ecologically unique, unmatched in the natural world. Baikal holds an enviable list of superlative titles. As mentioned, it is the oldest, deepest, and most voluminous lake on the planet. Its water is among the purest in the world. It is home to thousands of

Figure 2.2 Two women converse on a bench by Lake Baikal. Photo Credit: Author

endemic species. The lake is a source of local legend and historical lore, a defin-itive aspect of Eastern Siberia (Fig 2.2).

Baikal was formed approximately twenty-five to thirty-five million years ago when the Indian subcontinent collided with Asia. The impact created a *graben*—a split between two fault lines—fracturing the earth just north of Mongolia. The same process that gave rise to the Himalayan Mountains also brought forth Lake Baikal. And just as the Himalayans continue to rise, so does Baikal con-tinue to widen, growing at a rate of two centimeters per year. However, it is not the width of Baikal that astounds, but rather its depth. Its present depth reaches more than two kilometers to its sediment floor—the deepest lake on the Earth. Scientists estimate that the sediment alone continues another seven kilometers beneath this. The astonishing depth of Lake Baikal allows it to pool water from more than 330 inflowing streams and rivers, rendering it the most voluminous lake on Earth. Twenty percent of all the Earth's unfrozen surface freshwater can be found in Lake Baikal. It is a freshwater reservoir of global proportions.

Not only is Baikal shockingly deep, it has still another surprising secret that is not repeated elsewhere on the planet: Baikal's water is oxygenated all the way to its maximum depth and supports life even at the lake floor, some 5,387 feet below the surface. No other deep lake can make such a claim. In other deep lakes, water below 650 feet is "dead"—it lacks oxygen and cannot support life. Baikal defies this rule. While scientists have some theories to explain the hospitality of Baikal's depths, no one can say conclusively the answer to this geological, biological, and hydrological mystery.[4]

The secret of Baikal's purity can be found in one of its astonishing array of endemic species, which number over four thousand (Galazii 2012). The *epischura baikalensis* is a microscopic shrimp that lives exclusively in Baikal. The lake holds literally billions of these tiny creatures who filter-feed for organic matter and, in so doing, remove impurities from the water. The power of these countless miniscule crustaceans, and the speed at which they feast on organic matter, defies expectation: a carcass of a dead cow thrown into the lake would be reduced to a skeleton in a week. The result is a water purity that is almost unmatched. Frozen Baikal water is as transparent as glass. The lake's water is so clean that it is not even recommended for use as drinking water: it lacks the vital minerals that human drinking water normally contains.

While the water of Baikal is miraculously clean, the biota that enacts this miracle is decidedly not. Any harmful toxins, the effluvia of modern industry and development, that find their way into Baikal's watershed are quickly consumed by the epischura, and other microorganisms, and subsequently conveyed into the food chain. As bigger fish eat the smaller fish, in increasing quantities, pollutants build up in the bodies of the animals. The health effect of these accumulated pollutants becomes especially consequential in Baikal's other famous endemic denizens: the *omul'*, the *golumyanka*, and the *nerpa*.

Omul' is a relative of the salmon and is a Baikal delicacy (Breyfogle 2013). *Omul'* is a "must-eat" for pescaterian tourists to the region. Train stops and bus stops are host to kerchiefed *babushki* hocking their family's catch—steamed or smoked. Fishing for *omul'* is regulated, although poaching is still a common occurrence. Locals fish for personal consumption, as one dimension of subsistence homesteading, and many also fish commercially for the resale value. Because of the heavy human consumption of *omul'*, and other Baikal fish, the toxins and industrial effluvia that find their way into the lake, and subsequently into the food chain, become problematic not only for the environment but also for human health.

Golumyanka, another important Baikal endemic, is a highly unusual fish in many respects. It lives throughout the entire depth of Lake Baikal—even a mile below the surface. The fish can survive this intensity of water pressure because it has no swim bladder. Also, its body is made almost entirely of fat. The fish is so translucent that a person can read newsprint through its body. Local fisherman claim that if you bring the fish to the surface and leave it in the sun, it will melt into a puddle of oil. The *golumyanka* is also one of the few fishes in the world that gives birth to live young. This fatty, deep-swimming fish is the staple diet of the *nerpa*, and the toxins that build up through its

Figure 2.3 A souvenir stand at a Listvyanka street market with nerpa stuffed animals, magnets and figurines. Photo Credit: Author

own feeding get passed along in this manner to the lake's largest endemic predator.

The Baikal seal, known locally as the *nerpa*, is a freshwater seal that is the virtual emblem of Lake Baikal. The *nerpa* is a close relative of the northern ringed seal, but it resides in a freshwater lake that is literally hundreds of miles from the nearest ocean. No one knows for sure how the *nerpa* arrived in Lake Baikal, but it is believed that their ancestors swam up the Yenisei and Angara rivers from the Arctic Sea—a journey of some three thousand miles. Now permanently ensconced, the *nerpa* are a veritable totem for Lake Baikal. Their face and likeness adorn souvenir stands: embroidered onto tee shirts, molded out of clay, stuffed as plush toys (Fig. 2.3). Seldom seen in the wild, there are two "Sea World" style aquariums near the lake where tourists can watch trained *nerpa* frolic and play. While they are still hunted for meat and fur, the seals are also protected and their populations are regulated by the government. But the effects of built-up toxins in seal blubber has been a greater worry for the species in recent decades than the threat of hunting. Toxins are thought to have caused a massive die-off of *nerpa* in 1986. With immune systems weakened by toxins, hundreds of these unique animals succumbed to an epidemic in canine distemper that year. The epidemic helped establish

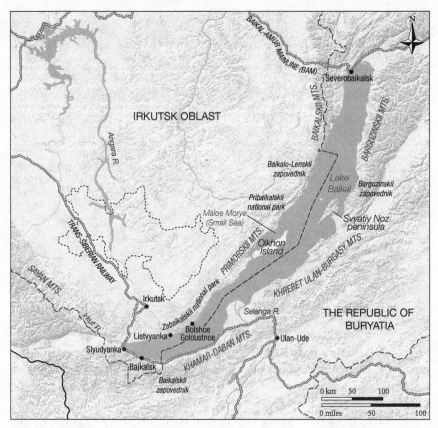

Map 2.1   Map of Baikal and surrounding territory.

the *nerpa* not only as the face of Baikal but also as the frequent face of Baikal environmental activism.

These are the most famous endemic residents of Lake Baikal, but they join thousands of others, some of which are endangered and many of which are listed in the Red Book, the official government document on rare and endangered species within the Russian Federation. These endemic species and their non-endemic neighbors join with the human population to create the biota of Lake Baikal.

Baikal and Irkutsk: Geography, History, Governance, Economy, and People

The lake, which is 395 miles in length, lies on the border between two regional jurisdictions: the Irkutsk *oblast* holds the western shore, and the Republic of Buryatia lies to the east. The lake is walled by mountains on all sides: the

Primorskii and Baikalskii ranges line the Irkutsk shore of the lake, while the Barguzinskii and Khrebet Ulan-Burgasy ranges rise to the east in Buryatia. The Khamar-Daban range flanks the southern shore. The Baikal basin contains several different ecological terrains. Much of the surrounding territory is *taiga*: boreal forests of coniferous and deciduous trees. The western shore also has large stretches of *steppe*, or grassland plains. Given the widening tectonic rift and its surrounding mountain ranges, the Baikal region is seismic and prone to earthquakes which vary from unnoticeable to those of such scale that entire swaths of land collapse into the lake. Hot springs also bubble up, particularly in the north and south.

The lake is fed by more than 330 inflowing rivers and streams, but one tributary, the massive Selenga River, provides more than half of its water. The Selenga delta, located in Buryatia on the lake's eastern shore, stretches out over twenty-five miles, and it is an important breeding ground for many of the lake's fish and birds. There is only one river that flows *out* of Baikal: the Angara. The Angara flows westward, eventually joining the Yenesei River and continuing on to the Arctic Sea. Irkutsk, the capital city of the Irkutsk *oblast*, sits at the confluence of the Angara and Irkut rivers. It is at this point that the first of the Angara's six hydroelectric dams is located. The dam powers the city of Irkutsk, and its construction, completed in 1959, raised the water level of the Angara corridor by several meters, flooding a stretch of the Trans-Siberian Railway, which had to be re-routed from Irkutsk to Slyudyanka at the tip of Lake Baikal.

Much of the territory around the lake is unsettled and inaccessible by road, but there are several cities and villages that dot its shores. The two largest cities near the lake are Irkutsk and Ulan-Ude, the capitals of Irkutsk *oblast* and the Republic of Buryatia respectively. Irkutsk has a population of approximately 590,000 and Ulan-Ude is home to approximately 400,000. These capitals are located 70 and 140 kilometers from the lake itself to the west and to the east. There are three smaller cities that sit on the lakeshore: Slyudyanka, Baikalsk, and Severobaikalsk. The first is a mining town and a regular stop on the Trans-Siberian Railway. The latter two were built to support major Soviet industrial projects. Baikalsk was created around a cellulose plant that was constructed in the early 1960s, about which more will be said later in the chapter. Its current population is about 14,000. Severobaikalsk, home to approximately 25,000 people, lies at the far northern tip of the lake. This city was built to support the construction of the Baikal-Amur Mainline (BAM), a spur off the Trans-Siberian Railroad (Ward 2009). A number of other towns and villages dot the shoreline, particularly in the southern portion of the lake. Previously, in the Soviet Union, villagers found employment in forestry, collective farms, mining, and factory work. These villages now survive mostly on fishing, domestic agriculture, and tourism.

Despite these pockets of human civilization, most of the territory surrounding Lake Baikal is protected. The country's first strict nature preserve, the Barguzinsky *zapovednik*, was created in the lake's northeastern territory in 1917 to protect the endemic Baikal sable from extinction (Weiner 1988). The Baikalskii *zapovednik* was established in 1969 on the lake's southeastern shore. The Pribaikalskii National Park and the Zabaikalskii National Park were established in Irkutsk and Buryatia, respectively, in 1986 (Roe 2016). North of the Pribaikalskii National Park sits the Baikalo-Lenskii *zapovednik*, which was also established in 1986. The Russian State Forest Service holds much of the remaining territory in the Baikal basin.

Although the lake sits between two different regional governing bodies, the gateway to Lake Baikal for most visitors is the city of Irkutsk. Founded in 1686, Irkutsk began its life as a military outpost and trading center. The Russian Empire would collect sable fur as tribute from the native Buryat people and trade goods from China such as tea, silk, precious metals, and stones. All these items were ferried through Irkutsk. The first road between Moscow and Irkutsk was completed in 1760, and this cemented the importance of the city in commercial trading. The city became home to wealthy merchants, who sought to establish for themselves opulent lifestyles to mimic the aristocracy. In 1821, the city became the seat of the governor-general for all of Eastern Siberia, which included Russia's colonies in North America. Irkutsk was home to political authorities whose sway spanned continents and who also had ambition to bring culture and learning to their resident city.

Arguably the greatest influence on the city's character came from the introduction of political exiles following the Decembrist revolt in 1825 (Diment and Slezkine 1993; Naumov 2006). When a segment of Russian nobility refused allegiance to the new tsar, Nicholas I, instead demanding a constitutional monarchy, suffrage, land reform, and freedom for the serfs, their rebellion was quickly put down. Five of the rebels were hanged, and over 130 were exiled to Siberia, Kazakhstan, and the Far East. After their terms of hard labor had expired, several of these fallen aristocrats set up homesteads in Irkutsk. Although the tsar gave them permission to declare themselves widows and remarry, the wives of the Decembrists chose to cede their titles, lands, and privileges and follow their husbands into exile. The presence of St. Petersburg's high society in this distant Siberian city helped Irkutsk flourish into a center for culture and learning. They established schools, hospitals, and theaters for the local population. The houses of the Trubetskois and the Volkonskys served as social and cultural centers, hosting various learned societies, musical concerts, and symposia. The Decembrists brought an intellectual, cultural, and political elite to Irkutsk, and indelibly altered the city's character.

These three streams—the wealthy merchants, the exiled nobility, and the central administration of the governor-general—combined to create in Irkutsk a beacon of European civilization that amazed the visitors who passed through it. Grand architecture joined ornate traditional Siberian wooden homes to give Irkutsk a hybrid look of a European capital and a homey hinterland. Called "the Paris of Siberia," the city was replete with a telegraph, train station, dramatic theater, and learned societies. This reputation continued into the Soviet era, which saw the city grow as a center for higher learning as well as for large-scale industrial development. Like the rest of Russia, Irkutsk was behind the Iron Curtain, but it had a reputation for free-thinking and independence (Diment and Slezkine 1993), two traits that were eventually borne out in the battle over Baikal.

Irkutsk is an *oblast*, or province, within the Russian Federation. Like all other subject territories, the Irkutsk *oblast* sends two delegates to the Federal Assembly. It is part of the Siberian Federal District, which is overseen by a presidentially appointed administrator who ensures local compliance with federal law. The head of the *oblast* government is the governor, a position that was popularly elected until 2004 when it was changed to a presidential appointment. The Legislative Assembly for the *oblast* writes laws for the region and ensures compliance. Within the *oblast* are a number of smaller administrative bodies governing districts and cities. The city of Irkutsk has a city council with five deputies. The administrative head of the city government is the mayor, which was popularly elected until 2015, when it was changed to a position appointed by the city council.

The Irkutsk *oblast* is home to 156 different ethnicities, but the vast majority (88 percent) is ethnically Russian. The second largest ethnic group is the Buryat people, who comprise slightly more than 3 percent. Although the Russian language and culture dominate, themes from indigenous Buryat culture are honored and do survive. The name "Baikal" comes from the Buryat words meaning "rich lake." The shaman religion of the Buryat people also infuses the region with its traditions and legends.

As the official state religion, Russian Orthodoxy claims the greatest number of adherents to its profession of faith. Seventy-three percent of Irkutsk residents are Orthodox believers. The region is also home to a surprising number of Catholics. The Polish, German, and Armenian prisoners exiled to Siberia in previous centuries brought their faith with them, and the Irkutsk *oblast* has thirteen Catholic churches. The Irkutsk *oblast* is also home to many Buddhists. There are six datsans in the region. The Russian center for Buddhism is located in neighboring Ulan-Ude. There is a Jewish presence in Irkutsk and a working synagogue. Muslims compose 6 percent of the Irkutsk *oblast*, which has four mosques. The city of Irkutsk is also home to a number of non-traditional religions, including Hare Krishna and the Latter-Day Saints.

The economy of the Irkutsk *oblast* is mixed. Industry accounts for 28 percent of gross regional product (GRP), with large enterprises involved in metallurgy; hydroelectric power generation; aviation manufacturing; large truck manufacturing; pulp, paper, and printing; chemicals; food production; and the extraction and refining of oil, gas, gold, and coal. Agriculture accounts for approximately 6.5 percent of GRP. Eighty-five percent of the Irkutsk *oblast* is covered in forest; the region holds 12–13 percent of all Russia's timber stock (Blam, Carlsson, and Olsson 1998). Tourism is a small but rapidly growing sector of the economy, employing 4 percent of residents in the city of Irkutsk. But the city's reputation as a center of higher learning remains its hallmark. At 14.2 percent of the population, more individuals work in the sphere of education in Irkutsk than any other sector of the economy (Irkutskstat 2014).

This high proportion of scholars and students, clustered together in the city of Irkutsk, provided a critical mass for activism. The locale was a crucible, melding higher learning, diversity, and a legacy of political dissidence: a crucial combination for expanding the "prospectus of activism" (Brown 2016). All of these factors coalesced in Irkutsk to enable local environmentalists to accomplish a feat that no other group in the Soviet Union had successfully done: to spontaneously and publicly speak out against the government with the aim to redirect planned industrial development. They became an environmental civil society.

Politics and Civil Society in Russia

The energy of environmental activism in Irkutsk stands in contrast to attitudes and politics in Russia more generally. As the Soviet Union dissolved and Russia moved to adopt a liberal democratic form of government, Western observers eagerly anticipated the development of a thriving civil society that would lead its country toward a bright future (Starr 1988; Hosking 1991). It was a hope that failed to materialize. Instead, scholars soon bemoaned a "Soviet legacy" that caused the country to lag behind its Western counterparts in measures of democratization, public engagement, and social trust (Howard 2003; LaPorte and Lussier 2011).

And yet, the apparent fatalism with which contemporary Russian citizens view their social and political circumstances, and their general apathy toward citizen involvement, is far from baseless. It is a political culture borne of centuries of autocratic rule and thwarted attempts at social change, flowing in a more or less continuous stream from the sixteenth century through the present. An absolute monarchy, the Russian Empire was the last of Europe's autocracies to fall.

Over the centuries, it had crushed the periodic uprisings that threatened it, and the Russian tsars successfully resisted all attempts to open the door to a more representative government.[5] This history granted Russia a kind of exceptionalism in Europe, and it set the general pattern by which the subjects of autocracy learned the limits of their public endeavors and civil strivings.

Neither was the subsequent Soviet regime much different in respect to citizen involvement. Rather than having itself constituted by the people, the Communist Party sought to remake the public in its own image (Bauer 1952) and to integrate itself even more completely into citizens' daily lives than the absolutist monarchy of the tsars (Lewin 1985). The Soviet state was not static across its seven decades of rule—it vacillated between deadly purges and relative thaws—but it consistently forbade social activity that was not aligned with some organ of the Communist Party. Through its system of censors, the Party dictated the only proper interpretation of what was desirable or undesirable, permitted or verboten. While activist-minded citizens could find a place for themselves in the Komsomol or in a variety of Party-sanctioned social endeavors, the general public felt compelled by the Party into "volunteerism" quite involuntarily (Chase 1989). Mandated volunteer workdays, or *subbotniki*, were resented as no more than uncompensated labor. Soviet citizens retreated into private life, offering only kitchen table critique (Brodsky 1986) and sardonic satire (Yurchak 2006) to affirm the separation of self and state.

In 1986, the Soviet Union began the process of political change that led to its eventual demise and set the stage for its subsequent development. The era proved to be extremely tumultuous, and the public was thrown from one ideological extreme to the other, only to be pulled back again. The transition from the Soviet Union to the present can be divided roughly into three periods, characterized by differing trends: *perestroika*, shock therapy, and Putinism.

PERESTROIKA

By the time Mikhail Gorbachev became Secretary General of the Communist Party in 1985, a large number of problems that had been simmering in the background of Soviet society—from corruption, to economic stagnation, to military breakdown, to environmental degradation—came suddenly to the fore. Gorbachev shocked the nation by initiating his program of *perestroika*, or "restructuring," to address the problems in the fraying Soviet system. A critical component of this plan was his call for *glasnost*—openness and transparency—in a society that had been controlled by censors for decades. For the first time in modern memory, critics could vocalize their dissent without reprisal, political prisoners were pardoned, corrupt officials were exposed and removed, and the press finally published what had previously been repressed (Hosking 2001; see also Yurchak 2006). The people were allowed some freedom of association and the right to independent protest. What had been a Party-dominated civil society or a loose connection of unorganized concerned citizens could finally achieve association, autonomy, and voice.

At the same time, Gorbachev's economic reforms and moves toward decentralization had the unintended consequence of creating a crisis in production. Scarcity was rampant, and the federal fiscal situation worsened. But as the controlling arm of the state was loosened, the demands by the people grew louder, and the political-economic apparatus was no longer equipped to meet them. Although intended as a slow reform movement, *perestroika* precipitated the end of the Soviet Union,[6] which ushered in still more radical change.

SHOCK THERAPY

The end of the Soviet Union was heralded in the West as the triumph of liberal democracy and capitalism. It empowered a growing ideology— neoliberalism—that repudiated state intervention except insofar as it fostered free-market mechanisms. When Western experts advised Russia's new leaders on social and economic policy, they helped infuse the post-Soviet transition with neoliberal assumptions—and they had an eager audience among the anti-Soviet reformers (Appel 2004). The result was the largest transfer of ownership from public to private hands ever attempted in modern history. Called "shock therapy," the program called for immediate freedom from price controls and near total privatization of productive enterprises. The result was a catastrophic economic collapse.

The theory behind shock therapy was that it would be the quickest route from a command economy to a functional market-based one. But the sudden market reforms were rolled out without sufficiently strong state institutions to enforce rule of law (Hamm, King, and Struckler 2012). The owners of the new privatization "vouchers" were ignorant as to their value, which allowed a small group to rapidly consolidate the Soviet industrial infrastructure under their control. The explosion in wealth inequality that resulted gave birth to the Russian "oligarchs." As the state fell further into bankruptcy and debt, it left an institutional void at a time already troubled by massive social upheaval. The results of this confluence of crises were hyperinflation, payment arrears, unemployment, asset stripping, rampant corruption, swindles and pyramid schemes, and two devastating currency crises in 1992 and 1998 that caused most people's savings to evaporate. Mafia ruled the streets, the standard of living plummeted, and many people were plunged into poverty, while a small cadre became wealthy beyond what could have even been imagined just a few years before (Stiglitz 2002).

Although the post-Soviet years witnessed an explosion of political opportunities, the resultant chaos only furthered the sense of instability and powerlessness in the general public. Instead of one political party, there were literally hundreds. The devastating economic reforms of privatization and "shock therapy" were imposed unilaterally by Russia's president, Boris Yeltsin, and criticism erupted in parliament as the effects played out. The crisis precipitated a power struggle between the presidency and the parliament, with Yeltsin finally disbanding the legislative branch completely and calling for new elections and a new constitution that granted greater powers to the executive (Associated Press 2007). Then, facing a competitive election bid in 1996, Yeltsin secretly traded state assets for campaign donations from oligarchs in a scheme known as "loans-for-shares" (Stanley 1996). Not only was the economy in turmoil and society upended, the government itself was loath to abide by the rule of law. While the beginning of the 1990s showed high public support for democracy, by the end of the decade the desire for order had become dominant (Pew Research Center 2012).

PUTINISM

Vladimir Putin assumed the presidency of the Russian Federation at midnight on January 1, 2000. He had been appointed first deputy prime minister mere months beforehand, and seemingly out of nowhere assumed command when Yeltsin unexpectedly resigned. Thus, he went into his first election three months later with the status of an incumbent and won 53 percent of the vote. He gained the popularity of a wartime president through his tough stance on Chechnya's separatists, whose tactics included domestic terrorism. Two large centrist political parties merged to form United Russia, giving Putin a majority in the legislature and a political party to move his agenda forward. More importantly, he began the process of stabilizing Russia's economy and reasserting the authority of the state in what had devolved into near anarchy.

Putin quickly stabilized the economy by repudiating the neoliberal reforms of the 1990s and returning authority to a state that became decidedly interventionist. He reconsolidated the state's hold over the oil and gas industry, which provided more than half of federal revenue (Sixsmith 2010). He instituted a flat tax, which brought the rampant tax evasion under control (Gorodnichenko, Martinez-Vazquez, and Peter 2008). He brought charges against several oligarchs, which, although politically motivated, were justified in the public mind by the thievery and extreme inequality of the 1990s. The economy stabilized and Russia began to finally move forward.

But while Putin was credited with steering the country out of chaos, his governance (alternating as president and prime minister) has also been characterized by a growing return to authoritarian rule. In the guise of prosecuting oligarchs for tax evasion and other crimes, the independent media was decimated and mostly returned to the hands of the state (Lipman 2009). In the name of efficiency, popularly elected positions were changed to presidential appointments. Rules and requirements for political parties were repeatedly altered to minimize any opposition (Henry 2006), elections were increasingly suspect, and by 2007 United Russia could count on regular wins of above 50 percent of votes. By the time of Putin's return to the presidency in 2012, allegations of voter intimidation and fraud had reached significant proportions, and independent election monitors refused to uphold the validity of Russian elections (Mendelson 2008; Barry and Schwirtz 2011; Enikolopov et al. 2013). Meanwhile, the Kremlin began rebuilding a decidedly anti-Western ideological apparatus comprised of Russian nationalism, religious Orthodoxy, and hypermasculinity (Kaylan 2014). The concomitant cult of personality around a political figure who has not left national leadership for sixteen years has led many commentators to dub the new regime: "Putinism" (Zakaria 2014 ; Laqueur 2015).

For all its uncertainty, *perestroika* allowed an opening for citizen involvement. Environmental groups, who were seen as politically benign, were the first to take advantage of the opportunity. Environmental public meetings and rallies saw participation figures in the tens of thousands (Green 1990). In 1989, the Socio-Ecological Union mobilized the first nationwide protest that halted construction of a planned canal (Weiner 1999). During this time, environmental protest successfully closed or prevented the construction of more than fifty nuclear reactors, as well as a number of hydroelectric stations and gas pipelines (Henry 2006). But the 1990s decimated the burgeoning public sphere. Existential insecurity took precedence over citizen involvement, in a classic case of materialist versus post-materialist values (Inglehart 1995). And the chaotic 1990s with its rampant corruption fostered a climate of social distrust (Evans 2002; Domrin 2003; Jakobson and Sanovich 2010; Richter and Hatch 2013; Brown 2016). But it would be a mistake to overstate the level of citizen activism in the late Soviet years. *Any* independent activism would have been unprecedented. While post-Soviet political and economic circumstances have done little to encourage a strong civil society, its lackluster state is no historical anomaly in a country that has never shown robust citizen participation. The low levels of public engagement in Russia historically and into the present make the environmental activism that began in Irkutsk at the height of the Soviet government all the more surprising.

The History of Environmental Activism around Lake Baikal

Despite the many political conditions that have impeded civil society else-where in Soviet and post-Soviet Russia, the city of Irkutsk in the Baikal basin developed a strong, homegrown environmental activist community, buoyed by its belief in itself and the rightness of its cause. The public reverence for Baikal that extends throughout Russia is both a cause and a product of decades of struggle by regional environmentalists who have striven to protect the lake from anthropogenic harm. Although emerging in a very different social and political-economic context than the present, environmental activism in the Baikal basin during the Soviet period also took place within the field of power. The historical development of Soviet environmentalism is similarly dynamic as powers wax and wane. However, the field was largely monopolized by the Communist Party (Weiner 1999; Josephson et al. 2013), which sought to command all three forms of generalizable power: financial, legal, and civil.

Although there was dynamism during the Soviet period, that movement was constrained by the Party's near monopoly over the different types of generalizable power. The Communist Party was in charge of the state, with its rule-making authority, as well as production and commerce, which prevented concentrated financial power from accruing to social groups outside the party. The Party similarly forbade independent association. Instead, it attempted to produce ersatz civil power through Party-sanctioned civil society. The mandatory nature of public involvement limited the reach of general "worthiness" and curbed potentially threatening civil power from forming. There were outlets for active and motivated citizens and there were beneficial works being performed, but the civil power built from the virtue and worthiness of independent voluntary action was limited by the state's involvement. The Communist Party recognized the diversity of generalizable power sources and sought to dominate each and every one, closing off the field of power to outside players. The effects of field closure were made manifest in the increasingly deplorable environmental conditions that developed and steadily worsened over the decades of Soviet control (Pryde 1991; Feshbach and Friendly 1992; Josephson 2002, 2005; Josephson et al. 2013; Bruno 2016,). Soviet civil society around Baikal, for all its efforts, could not stem the tide until the Party's monopoly over the field was broken.

When Soviet industrial projects first began to encroach on the Baikal basin in the mid-1950s, the lake already held a significant sway on the public mind. The indigenous peoples who had lived in the region for centuries had a religious reverence for Baikal that provided a cultural foundation for its more general appreciation. The many research institutions in Irkutsk were steadily producing greater knowledge that confirmed the lake's ecological uniqueness and scientific importance. Moreover, the region already had faced the consequences of human overindulgence; in the early 1900s, the Baikal sable was hunted to near extinction for its fur. In response, the Imperial government established the first *zapovednik*,[7] or state nature preserve, on the shore of Lake Baikal in 1917 so as to protect and preserve the endangered sable. Baikal was considered both beautiful and sacred—it was a gift, something wondrous, with which technology and industry should not meddle.

When the long arm of Soviet development reached the shore of Baikal, it faced massive public opposition for the first time, making the lake the national symbol for nature protection. Writes historian Nicholas Breyfogle:

> From the late 1950s on, Lake Baikal became the centre of one of the most visible, powerful and successful environmental movements in the Soviet Union, which deeply influenced environmentalism throughout the USSR (and now Russia). Until Chernobyl' in 1986, and in many respects even after that too, Baikal was the environmental cause of the post-World War Two period. Others may have been worse in terms of their impact on the natural world (such as the draining of the Aral Sea or Chernobyl'), but none grabbed the Soviet imagination more strongly or deeply than Baikal. (Breyfogle 2015, 147)

Baikal represents a unique case in Soviet history where public opinion came out actively *against* the state. As historian Douglas Weiner writes:

> Unlike any previous environmental struggle, the fight to protect the vast lake from physical alteration and industrial pollution embraced not only the various branches of committed nature protection activists but a broader public as well. Such public participation imbued the struggle around the lake with a larger meaning as an incipient general protest against the rulers' abuses of power. (Weiner 1999, 356)

The first fight began in 1958 over increasing water flow out of the lake to better feed the Irkutsk hydroelectric station. It was only the first of many.

THE GRIGOROVICH PLAN

According to Buryat legend, Baikal has 330 sons but only one daughter, named Angara. She fell in love with Yenisei, and eloped with him. That is why 330 rivers flow into Baikal and only one, the Angara, flows out and eventually joins the Yenisei River, which continues on to the Arctic Sea. Baikal was so angry that Angara had run away, he threw a rock after her. "Shaman rock" stands at the mouth of the Angara, a few feet jutting out above the water's surface, scarcely visible, in the middle of the river.[8]

The Angara joins the Irkut River at the city of Irkutsk. The Soviet government built the first dam on the Angara at Irkutsk in 1950 to provide power to the city. N. A. Grigorovich was the chief engineer of the Angara Sector of *Girdroproekt*, the hydrological planning commission. In 1958, he proposed to detonate an explosion at the mouth of the Angara River in order to increase the water flow to the hydroelectric dam in Irkutsk. The proposed explosion— intended to be 50 percent larger than at Hiroshima (Weiner 1999, 357)— would lower the water level in Baikal by several meters.

Soviet history is rife with "projects of the century" (Josephson 1995) whereby humanity would show its mastery over nature: diverting rivers, creating canals, building atomic energy stations, and so forth. Often these would result in catastrophic destruction of human life and natural ecosystems. Geo-engineering Baikal proved to be the limit to Soviet attempts at "mastery over nature." In Irkutsk, with its rich regional culture and history, there was a population primed to speak out against the proposal.

Chief among the local campaigners was Grigorii Ivanovich Galazii, head of the Limnological Institute at the Siberian Academy of Sciences, who would, decades later, be elected to the Supreme Soviet and pass the Law of Baikal, enshrining its protection at the highest level of government. In these days, however, he was a young scientist who was passionate about Lake Baikal and the delicate balance its ecosystem maintained. Galazii joined two other local writers who were against the project in crafting a letter to *Literaturnaya Gazeta*, a major national newspaper. Their letter, "In Defense of Baikal," described the potential negative environmental impact of the Grigorovich plan. Within a month, a thousand letters of support came in from across the Soviet Union (Weiner 1999). Meetings were held, and conferences were organized; the outpouring of public sentiment against a planned industrial project was unprecedented in the Soviet Union. The virtue of these men of science and letters, speaking out on behalf of a unique national treasure, motivated and mobilized the masses. The state was caught unawares, but neither was it deeply invested in Grigorovich's proposal. The plan was scrapped. In this case, the environmentalists had won. However, this was only the beginning in the public fight for Baikal.

THE BAIKALSK PULP AND PAPER MILL

Later in the same year, the Soviet government announced its intention to construct a cellulose plant on the shore of Lake Baikal. The Baikalsk Pulp and Paper Mill (BTsBK),[9] as it would eventually be called, became a lightning rod for environmental concern in the region across the entirely of its existence (Yanitsky 2011). The two commissioned plants, BTsBK and the Selenginsk Pulp and Paper Mill to be built on Baikal's largest tributary, the Selenga River, were bound up in the Soviet military-industrial complex. In the 1950s, tires for fighter planes and bombers were made of cellulose cord. To produce high-grade cellulose, plants needed large quantities of timber and water. Other lakes in the Soviet Union had already been clear cut, but the Baikal basin was still flourishing with timber. Moreover, the clean, pure water of Baikal was argued to be necessary to produce the highest quality cellulose, and therefore the best aircraft tires.

Again, the outcry against the proposed plant was enormous. The all-union presses picked up the story and helped foment widespread dissent. Pressure was so intense that the Soviet state felt compelled to respond to citizen concerns. In April 1960, the USSR Council of Ministers adopted a law on water pollution that would require the latest pollution abatement technologies in all new factories. Then on May 9, the RSFSR went further and passed a law delaying the start-up of the Baikalsk and Selenginsk factories until waste purification systems were put in place.

But even these concessions could not quell the outrage. Letter writers reminded the public that Baikal is seismic; even the most advanced water purification system would be irrelevant if the factory faced a major earthquake. Meanwhile, as the fight raged on, the product that the mill was designed to produce became increasingly obsolete. Nylon and polyethylene cord replaced cellulose as the industry standard. Gone along with it was the necessity for extra-pure Baikal water for military-grade cellulose. Instead, BTsBK would make paper and pulp—as it could anywhere else—only to put Baikal at risk. Repeated public letters, proclamations and conference proceedings reiterated opposition to the mill, with long lists of signatories, including respected scientists and various professional associations.

There were countervailing opinions as well: workers were beginning to flock to the small fishing village that would eventually become the city of Baikalsk. There were jobs to be had in the construction of the factory and building the city that would surround and support it. Eventually there would be jobs in the mill itself. Claims were made on behalf of the mill from a developmentalist standpoint, arguing that the economic benefit would outweigh the potential harm, and that the sheer size of Baikal would quickly dissipate the pollutants from the mill. But these did not outweigh the increasingly aggressive stance of Soviet scientists and others who were against the development. Baikal would be their line in the sand.

As the mill's opening date grew closer, the flurry of activity reached a fever pitch. The Moscow Society of Naturalists held a major conference, replete with militant resolutions. Once again, the national newspapers joined the fight. Yet, for all the action, this battle was lost, and the Baikalsk Pulp and Paper Mill began operation in 1966. It is in the conflict over the Baikalsk paper mill that field of power becomes central for explaining why the existence of a form of civil society in the Soviet Union, or why documenting the presence of multiple, active, interpretive subjectivities in a polity, is not the end of the analytic road. When its monopoly on the field of power fractured with the emergence of independently motivated environmental activism to protect Baikal, the Communist Party responded with a power play, using its command over legal power to undermine and repress burgeoning civil power.

REPRESSION OF BAIKAL ACTIVISM

The state had its mill, and environmentalists succeeded in pressuring the government to build it with the highest pollution abatement techniques of the time.[10] But critical voices continued to speak out against the mill and other industrial encroachment in the Baikal watershed. The state was growing impatient with its green patriots and took measures to stifle the dissent. Censorship of the media was simply standard operating procedure in the Soviet Union, and it is widely suspected that criticism of the mill was eventually suppressed in the national media. After chasing the story regularly for some time, national news outlets stopped mentioning the paper mill at Lake Baikal. Articles on the mill were difficult to find for a number of years, at first in the late Khrushchev period, and again under Brezhnev. There was a brief resurgence in critical publications during the power changeover, suggesting some editorial flexibility in the political transition, followed by a second clampdown. While historians have not found a "smoking gun" that confirms an official gag order on the mill, there is reason to suspect that limitations on Baikal protest—at least at the national level—were in place.[11]

However, local presses and scientific venues remained relatively open to local dissent, and those passionate about protecting Baikal still insisted that they be heard. Grigori Galazii was one who would not keep quiet about the fragility of Baikal's unique ecosystem and his deep concern that human damage would eventually render the ecosystem unrecoverable. Galazii was, by this point, the head of the Limnological Institute, a prestigious research center affiliated with the Soviet Academy of Sciences. It had a large laboratory, and Galazii used it to produce scientific measures of pollution in the lake and to demonstrate its negative consequences for the organisms therein. It was pure science, but it produced inconvenient facts for the administration. Galazii could not be fired from his position, but he could be "re-distributed." The government relegated Galazii to a single room, without a laboratory, that they dubbed the "Baikal Museum," and placed him in charge. Without access to a significant laboratory and a team of researchers as he had previously had at the Limnological Institute, Galazii could not produce the studies that were implicating industry in Baikal's degradation. He continued to speak out for Baikal, but the illustrious title of the Academy of Sciences was no longer appended to his name.

The generalizable power that comes from determining the rules can be deployed to diminish other generalizable powers, and here the Soviet state ruled away the civil power accruing to the Baikal protection movement. It legally limited the reach of their call and retaliated against their leadership. It was not sufficient to end Baikal protection activism and activity, but the spread and the militancy of the movement was diminished, until the field of power again opened in the 1980s.

SOVIET ENVIRONMENTAL CIVIL SOCIETY: VOOP

In addition to the spontaneous social outpourings discussed above, which arose in response to an immediate threat, Irkutsk also had the same stable environmental civil society organization that existed throughout the Soviet Union. The All-Soviet Society for Nature Protection (VOOP) was an official organization, under the auspices of the Communist Party, and it collected the common efforts of scientists, teachers, naturalists, and nature *aficionados* to press for environmental protection. They were able to persist, despite state suppression of other forms of activism, because of the apparently "apolitical" appearance of nature protection (Weiner 1999). The Irkutsk branch of VOOP carried the banner locally. A brief investigation into the organization in the 1980s can give a sense of what local environmental civil society looked like around Lake Baikal during the Soviet Union.

VOOP was a Party organization and, as such, was modeled on the organizational culture within the Party itself. It had five-year plans with targets for the number of members, the number of raids conducted, the number of school activities organized, and so forth. Outcomes were not set by any particular improvement in environmental conditions, but simply by output of activity. The activities planned and conducted by VOOP were broad—ranging from placard competitions to workshops on preventing forest fires. In content, the message was similar to those of contemporary environmentalists. They believed nature enriches the human spirit and should be preserved for future generations. But they were also clearly aware that the state placed a priority on production, and they were not to question that. Instead, they emphasized "rational use" of nature's resources and the need to preserve them for the sake of more efficient productivity.[12]

VOOP members would sing the praises of the state in their official proclamations, conferences, and reports. Sometimes the ideology opened the door to discussions of nature protection. But more often than not, the need to pander to the ruling elites bordered on absurd. Conference speakers would begin a speech with paeans to the Soviet government, for example: "Questions of environmental protection have stood at the center of attention of the Soviet government since its first days of existence" (VOOP Archive, 1980, Delo 510, p. 28). School children would take part in activities organized by VOOP with titles, including: "For a Lenin-like Relationship to Nature!," "To love nature is to love your motherland," and "To love nature as Lenin loved her." (VOOP Archive, 1980, Delo 337, pp. 24–26). During one meeting, a VOOP representative encouraged the committee to work harder to propagandize environmental protection in the media. Her colleague agreed, saying, "Not just propagandize it, but give it a Marxist-Leninist position. That way it will garner more [visibility]" (VOOP Archive, 1988, Delo 613, p. 40). Nature protection was only valid insofar as it worked in accordance with the Soviet state and its larger agenda.

The archives themselves do not inform us about the internal states of these Soviet activists. The extent to which these individuals actually believed that Lenin was the appropriate role model for nature protection can only be inferred, but is doubtful. Certainly the suggestion that tying into Marxism-Leninism would garner them greater visibility implies strategic intention. But such use of ideology in no way discounts the possibility of true conviction in the Soviet project as a whole. As Yurchak (2006) explains, the tropes of Communist discourse became a kind of ritualized performance in their own right, neither cynical nor sincere. But the practice still corralled the discourse of environmentalism within the paradigm of Soviet politics, policy, and production. It ensured that, no matter what VOOP said about the environment, the authority and agenda of the Party would remain inviolate.

The terms were set by the Communist Party, but within the terms, activists were free to follow their environmentalist conscience. Under the auspices of VOOP, myriad environmental actions took place. In putting on actions and promoting environmental ethics, VOOP received bountiful benefits from its position as a state-sanctioned social organization. First, the organization never hurt for membership. Soviet citizens were expected to sign up for a number of clubs and associations, and VOOP was a box on a list of such organizations that workers could easily tick. Membership was in the thousands, but these were "paper members," of whom nothing was required. Second, it had a steady stream of ready income for its activities. Dues for memberships in official organizations were automatically deducted from payroll. Eventually, in the post-Soviet years, nonprofits bemoaned the payroll deduction practice as having destroyed the culture of philanthropy among the Russian public, forcing independent civil society organizations to look abroad for funding.[13] But during the Soviet Union, nature protection activities had ready money for their activities. Third, the press was a willing supporter in its efforts to propagandize environmental protection. The largest regional paper, *Vostochno-Sibirskaya Pravda*, regularly offered VOOP a full page of its newspaper to fill with content related to environmental protection in a section called *"Rodnik"* [Well-spring]. Within a censored society, VOOP had a regular megaphone.[14]

In the study of social movements, some see the ability to garner resources that support one's activities and spread one's message as the critical determinant (McCarthy and Zald 1977). The Communist Party provided abundant resources for VOOP's activities, but it wielded these as a kind of power, designed to ensure that the message of environmental protection and endangerment was in accordance with its own. The Soviet political opportunity structure could make space for a certain kind of environmental civil society, as long as it was non-threatening to the Party's larger agenda.

Without independent power, VOOP was more of an outlet for activity than a force for social change. Members of VOOP occasionally expressed frustration at their own ineffectuality. As one member bemoaned, "The Factory Council of VOOP, its social-technical committee, has no influence on the answers to questions of environmental protection. Regardless of required state inspections, there is no organized regulation of toxic air pollution from auto-transport" (VOOP Archive, 1986, Delo 518, p. 33). The Irkutsk All-Soviet Society of Nature Protection played by the rules. They would praise the state, frame their work in accordance with ideology, and mimic the organizational culture of the Communist Party. Still they were denied influence on the decision-making powers.[15]

THE BAIKAL MOVEMENT

The mid-1980s witnessed a seismic shift in the field of power. *Perestroika*, first announced in 1985, opened up uncertainty and new possibility. The explosion at Chernobyl' in 1986 came to symbolize all the fear and frustration Soviet citizens felt for their own health and well-being in their increasingly toxic communities (Feshbach and Friendly 1992; Dalton et al. 1999; Weiner 1999; Petryna 2002). Political dissent was still novel and unpracticed, but environmental dissent was a banner the people felt they could safely hold high (Dawson 1996). The Party tried to align itself with growing public concern, and the proceedings of the 27th Congress in 1986 gave ideological cover for widespread environmental protest. When the state allowed free association for the first time in 1987, it was environmentalism that stood at the vanguard, ready to sally forth and onto the street. The first place where that opportunity was taken was Irkutsk.

Protecting Baikal remained a prominent concern in the public mind, despite the movement's retreat into abeyance following the construction of BTsBK. Deep frustration was simmering below the surface, ready to erupt again when the state loosened its reins during *perestroika*. The trigger for renewed environmental activism in Irkutsk turned out to be the *nerpa:* Baikal's unique, endemic freshwater seal. In 1986, *nerpa* began dying in droves. Animals would be seen heaving themselves upon the lake shore, sick and blinded. The diagnosis turned out to be an epidemic of canine distemper that had infected the seals and spread like brushfire. But, despite the official diagnosis, all eyes looked toward the paper mill as the chief culprit.

For decades, the mill had dumped its toxic effluvia into the lake, trusting in the water's volume and purity to absorb and disperse this waste. But these toxins entered the food chain and were found to have accumulated in *nerpa* blubber. Environmentalists claimed that toxic exposure had weakened the seals' immune systems, rendering them susceptible to distemper in epidemic proportions.

The state, meanwhile, had conceived a plan to construct a pipeline that would ferry waste out of the Baikalsk paper mill and into the Irkut River, where it would be carried, untreated, along the Angara, to the Yenisei, and eventually out to the Arctic Sea. Environmentalists deemed this "solution" to be even more dangerous than the current waste disposal reservoirs. Baikal is a seismic region; pipeline fractures and spills were the expected norm. Not only would this endanger Baikal, but it would taint the rivers from which Irkutsk residents drank, fished, and swam. The proposed plan to pipe the waste for dumping in the Irkut River was met with scorn and indignation. Members of the public decided to put their newly granted freedoms to the test.

A collective of active citizens began planning a protest march. They called themselves the "Baikal Movement" (*Baikalskoye Dvizheniye*). The leaders were drawn from the intelligentsia and the "workers' intelligentsia." They were active in the democratization movement during *perestroika*; generally they were the ones forming the new local elected councils in their communities. The Baikal Movement announced an independent protest march in October 1987 to prevent the BTsBK-Irkut pipeline. It proved to be the first independent protest in the Soviet Union following *perestroika*. One participant described it for me thus:

BAIKAL MOVEMENT PARTICIPANT: There was a large group gathered, and the KGB was standing by watching them. People were very uncertain—Are they really going to let us march? And so we took a

> few steps, and a few more steps, and no one was arrested! And so we
> just kept walking, all the way to [Kirov Square]!

And so marched the first picket in the USSR held independently of the
Communist Party. The protest was against the pipeline that would dump
BTsBK's waste into the Irkut River, but it was also, at its heart, a political pro-
test. As one observer described it to me: "The problem was environmental—
defending Baikal, but it criticized only one organization, the Central
Committee of the Communist Party of the Soviet Union." As *perestroika* pro-
gressed, questions of democratization overwhelmed environmental issues.
But environmental activism to protect Lake Baikal was midwife to these.

Leaders of the Baikal Movement generally moved into electoral politics.
When this happened they frequently abandoned environmentalism; it was
a green wave that they rode into elected office, but it was not a driving personal mission for them. But Grigorii Galazii's passion for Baikal never faded.
He was elected to the Supreme Soviet, where he pushed for a law to protect
his treasured lake, until he finally succeeded in passing the Law of Baikal on
May 1, 1999, as the first federal land-use regulation geared toward a specific
territory.[16] The law established boundaries, or "zones," around Baikal and
its watershed: a central environmental zone containing the lake itself and
its immediate shores, a buffer zone that includes the Baikal watershed,
and a zone of atmospheric influence that encompasses atmospheric circu-
lation bringing contaminants into the watershed. The law stipulated maximum allowable pollution levels in the central ecological zone. Despite
problems related to lack of oversight and vague requirements, the law has
remained the chief legal avenue by which activists challenge industrial pol-
luters through the present day.

BAIKAL WATCH

With *perestroika*, the Iron Curtain began to dissolve and the Soviet Union
allowed its borders to become more porous for people flowing out and in.
Environmentalists in the Soviet Union and the West sought out one another
and began to build transnational advocacy networks. One of the first eager
participants to reach out across this long-closed border was David Brower.

Brower was a legendary environmental activist in the twentieth-century
United States, the veritable creator of modern American environmentalism
(Turner 2015; Wyss 2016). Former president of the Sierra Club, Brower went
on to found Friends of the Earth, the League of Conservation Voters, and the
Earth Island Institute. In 1988, at the invitation of environmentalists, Brower
traveled to Moscow to attend a conference. He listened to reports on the state

of the environment in Russia, presented as lengthy and dry academic papers. Brower, a professional activist, would urge presenters to cut to the chase.

> GARY COOK: He would always ask them, "Well, what's the most important environmental problem? What is the place you want to save the most?" And with one voice they would always answer: Baikal.

In 1989 and again in 1990, Brower led teams from Earth Island Institute to Lake Baikal. In 1990 a group of twenty-five people came, toured the lake, and held a conference in Barguzinsky Reserve. They were impressed with the beauty of Baikal and all agreed that it should be targeted for preservation. Earth Island Institute created an offshoot organization, Baikal Watch, to serve this purpose, and Brower recruited a young activist named Gary Cook to lead it. Cook had been involved in the Earth Island Institute's campaign to protect dolphins from tuna nets. He held a PhD in economics from the University of California, Berkeley, and he had studied the Russian language. Traveling to Siberia for the first time with Baikal Watch at the end of the Soviet era, Cook was astounded by the catastrophic environmental conditions he encountered.

> GARY COOK: I traveled [to Russia] for the first time in late '80s, early '90s, and there were scenes there resembling Bosch: his vision of hell. There were cities that were so polluted, you'd come up a hill and you couldn't see the other side of the town from the top of the hill. We're talking two miles and it was basically black. [People] said it was always like that. For our poor, pathetic Western lungs it was like, "How am I going to survive this?!" That was a bit extreme, but the more places you went, more corners you turned, the more amazing scenes there were, [environmental] impacts you couldn't imagine. We were in the forest, and we knew we were in oil-drilling land. But we came over the hill and there would be a lake of oil that had been there for who knows how long. It wasn't a lake the size of Tahoe but it was a lake three to four miles across. Woah![17]

Baikal, by comparison, was actually quite clean. But industrial effluvia into the Selenga River and within the Baikal watershed were still significant concerns for environmentalists in the region. Baikal Watch saw its mission, not to solve these problems, but rather to teach Russians how to solve them, using Western civil society as the model. By 1991, the Soviet Union had disintegrated. The field of power was now open to the possibility of an independent civil society, holding and deploying its own unique form of power. And importantly, unlike the Baikal activism of the Soviet past, the new field of power would be globalized.

Conclusion

The Soviet era represents an instance in history when a social group—the Communist Party of the USSR—managed to establish a near monopoly over the field of power. Placing itself as the sole political party, fully in charge of all legal production and commerce, and outlawing public association other than Party-sanctioned activity, it succeeded in appropriating the principle forms of generalizable power in the modern era.[18] More so even than the autocrats who preceded them, the Communists sought to bring all public activity under the dictates of the Party (Lewin 1994). The oft-discussed theoretical questions about the extent to which the state *actually* controlled Soviet life misunderstand the Communist Party project. Explicitly or implicitly, the Party recognized that the state was only *one* form of power and sought instead to command them all.

The Communist Party's monopoly was never complete, and the field of power was flexible over time, leaving some opportunities for action. The Party proved susceptible to the sudden rise of civil power when a contingent in Irkutsk chose to speak up against the state in defense of Baikal. But the combined weight of military, economic, and political forces succeeded in constructing the Baikalsk Pulp and Paper Mill over the strenuous objection of a vocal public. State power was finally employed to cut off the remaining protest. However, throughout the ebb and flow of Baikal-based activism, there remained a relatively permanent civil society organization dedicated to environmental protection. It had extensive resources and regularly developed social activities to promote its mission. But dependent as it was upon the Communist Party, it remained fundamentally non-threatening to the field of power, and thus to the overarching agenda of its dominating player in the sphere of the environment.

As environmental problems compounded, there were no mechanisms by which the lay public could easily act to alter the course of events. Civil society as a power source capable of acting in the world was confined to such a constrained space of activity that it served mainly as a functionary of the commanding authority in the country: a dominating political elite. Essentially, it was so difficult for other agents to act in the field of power—even though people recognized problems, and there was a civil society working against them—the environmental projects of only one player could proceed, and the subsequent environmental problems compounded to the level of "ecocide" (Feshbach and Friendly 1992).

Seeing the field of power operate within the Soviet Union through the history of Baikal activism offers lessons that can apply to an analysis of contemporary conditions for civil society. The use of power plays to alter the scope of civil power is not a tactic limited to the Communist Party. Independent civil power remains a potentially threatening opponent to other power holders, and relationships between these power holders continue to evolve in this new and altered setting. As we look at Baikal environmentalism in the post-Soviet era—a period punctuated by neoliberal globalization and resurgent authoritarianism—the field of power helps to explain the similarities and the differences we encounter across the Soviet/post-Soviet divide, and it can illuminate the causes and consequences of more general trends in contemporary civil society around the globe.

3

Baikal Goes Global

The collapse of the Soviet Union was a seismic event. It shook the lives of the diverse millions stretching from Kaliningrad to Kamchatka, from the frozen Arctic to the deserts of Central Asia. It unsettled decades of entrenched geopolitical relationships. And, most importantly for our purposes, it unleashed a new arrangement for the field of power. The sudden creation of a privatized free-market economy allowed for the development of financial power, unleashed from state control. Meanwhile, Russia's civil society was buoyed by new freedoms of association and the ability to register official nonprofit organizations. Whereas the Communist Party had monopolized the Soviet field of power, the post-Soviet period allowed three distinct generalizable power holders to burgeon in the new social space.

However, these different powers exist relationally to one another in the larger field of power. Their individual development is closely bound to the dynamics of the others. When we consider the formation of civil society, for example, we must also understand how it exists in relation to other power holders. In the immediate aftermath of Soviet collapse, state power was thrown into disarray, and although financial power was newly freed from state control, the struggle to command it was volatile and often brutal. The instability and weakness of state and financial power left a window for the relatively free development of domestic civil society—an opportunity that would eventually be lost as their field of power opponents recovered.

The Soviet collapse also opened up Russian society, previously sequestered behind a metaphorical Iron Curtain, to the global plane, establishing a new spatial scale on which various power holders could participate. Transnational connectivity allowed each of the three power holders to augment their capabilities and created a new terrain for power interactions. Globalization compounds the field of power dynamics.

The dynamic development of local civil society around Lake Baikal demonstrates the changing form of the field of power. The history of three nonprofit organizations in Irkutsk shows the course of change. The first, Baikal

Environmental Wave (the Wave), is a nonprofit organization, officially incorporated in 1992, geared toward environmental education and advocacy work. The second, Tahoe-Baikal Institute (TBI), was formed as two sister organizations, one in the United States in 1989 and one in Russia in 1992, whose purpose was to carry out environmental-educational summer exchanges between American and Russian youth. Finally, the Great Baikal Trail (GBT), which formed in 2002, is a volunteer-based nonprofit that builds and maintains wilderness hiking trails in order to promote sustainable, low-impact ecotourism. While they are not the only environmental organizations in the region, they are among the most influential and the longest lived.

Together their stories illustrate the relationality of civil society to economic and political powers. When the latter are weak, civil society has the strength and independence to take a more aggressive and confrontational stance. In times when economic and state powers grow, civil society adapts as well and seeks apolitical, cooperative, and accommodational standpoints. These three histories also show the very different ways that access to the global can impact local civil society. Since the opening of Russia to the West, global relationships have become interwoven into local Baikal environmentalism so thoroughly that they can no longer be easily parsed. The global is a space that can be drawn upon to fill local gaps in knowledge, resources, and enthusiasm. As local groups cooperate with transnational partners, the very practice of global interaction becomes a magnet for local participation. These histories show a very different form of "transnational activism" than traditional understandings of the phenomenon would suggest. Rather than a coordinated campaign, the story of local environmentalism at Baikal paints a picture of activism engaging the transnational in multiple ways that generally (but not always) work to enhance civil society's power.

Origin of Baikal Environmental Wave

Jennie Sutton came to the Soviet Union in 1974 through an exchange program between Great Britain and the USSR for language instructors. While the program was no doubt designed to further the domestic interests of each country, such opportunities can also have powerful impacts on the lives of the individuals involved in the exchange. Such was the case for Sutton, who signed up for the program and was placed at the Institute of Foreign Languages in Irkutsk. She soon fell in love with her new home; she was fascinated by the Russian people and their culture. Although she was originally placed in Irkutsk for three years, she requested an extension, so as to complete a new textbook she was co-authoring. Eventually, she secured a position teaching and translating English at the National Academy of Sciences in Irkutsk. It was her position as a transnational actor at a major scientific research center that eventually enabled her to become a passionate and deeply committed environmental activist.

When discussing social movement participation, scholars often point to the "biographical availability" of individual activists: they must have the time to devote to their cause and little to personally risk in the process (McAdam 1986; see also Schussman and Soule 2005; Beyerlein and Hipp 2006). As such, the typical activist is young, unmarried, without children, and engaged in some form of flexible employment. In the experience of Jennie Sutton, I would add another dimension to her "biographical availability": she was positioned at the intersection of two different societies and was therefore "available" to receive input from one context that could be applied to the other. From a social network analysis perspective, Sutton was a node between two geographic networks. It was her position spanning two different national contexts that gave her the tools to become a leading environmental activist.

Although Sutton had always appreciated nature and the outdoors aesthetically, she came to environmental activism in her early 40s principally by virtue of her bi-national status.

> JENNIE SUTTON: I was aware that society was having an adverse impact on the environment, but, like most people, I had no idea to what extent this was happening until the end of the 1980s. I came across the Brundtland Report, which was the United Nations report on the state of the planet and of natural resources, and it happened to come into my hands because it was in English. I was teaching at the Academy of Sciences in those days, working with scientists on their English. I read [the Report], and that was probably in 1988. At that time, this was the height of *perestroika* here, which was an extremely interesting period.

Sutton felt, aptly, that she was living through history as she watched the changes brought by *perestroika* and *glasnost*. Environmental issues were tightly tied to emerging anti-authoritarian sentiment in the crumbling Soviet regime (c.f. Dawson 1996; Weiner 1999; Whitefield 2003; Rihoux and Rudig 2006). Sutton had her ear to the ground during these local environmental outpourings. She was working closely with scientists during the *nerpa* epidemic. She knew the scientist, politician, and Baikal environmentalist leader Grigorii Galazii personally, and she eventually translated his book on Baikal into English, at the request of Gary Cook from Baikal Watch (see Chapter 2). She paid close attention to the development of the Baikal Movement. She even marched in their demonstrations, although she did so only after asking the KGB for permission, so as not to risk her Soviet visa.

Still, it was not until a trip home to Britain that Sutton was inspired to devote herself fully to the environmental cause and to begin organizing the group that would eventually become Baikal Environmental Wave.

JENNIE SUTTON: In 1990, my mother was still alive, and I used to visit her in Britain every year. I just turned up there in May. It was after the Brundtland Report so the world community was becoming more and more aware of the facts, that [environmental problems] were on a different scale . . . that if we don't take steps to curb our use of natural resources and pollution, we are basically destroying what we depend upon—the planet.

While I was [in Britain], there was a series of television programs and films and discussions on these questions over a period of a whole week. I followed them avidly—I don't think I missed a thing. With the result [being] that I was convinced that this was Question Number One, that it needed to be taken seriously and that I had to take this message back to my colleagues here in Russia. So that is what I did.

I came back here . . . and in those days we had an English language club. We used to meet once a week and discuss things in English, as much as possible, and I said, "You know the situation is very serious. It is not just the Baikal region, it is the whole planet and we've got to turn our English club into an Eco-English club." . . . They accepted the information; some left, so some didn't continue in the English club. Some stayed, some people came and went. Over the next couple of years, some people came from outside [the Academy]—they weren't my students, but people who heard about us. Basically, the first couple of years, [there] were weekly meetings where we were, I would describe it as, basically [doing] self-education as to what was happening.

Among her students was Marina Rikhvanova, a young scientist working in the Limnological Institute. She was one of those who stayed in Sutton's English Club after the shift to "Eco-English." It proved to be a decision that would change the entire trajectory of her life. She eventually became a leader of Baikal Environmental Wave and developed her career in nonprofit environmental advocacy, winning the Goldman Prize and earning the ire of the Russian state.

MARINA RIKHVANOVA: I began studying English, and I really wanted to join Jennifer Sutton's group because she is British. And it happened that I was placed precisely in *her* group, and I was very pleased. But Jennie also led a club of English aficionados where we would discuss

various topics: poetry, writers, artists, and such. Then later Jennie
returned from England and said, "Everyone, from now on we will only
be discussing environmental issues. Who wants to?"

But it was not long before English fell to second place among those who decided
to stay in the group.

MARINA RIKHVANOVA: When studying environmental problems, we
immediately started to want to do something practical [to solve
them]. Once you are doing practical work, you want to discuss things
quickly . . . so we dropped speaking in English.

Leaving English behind, and branching out beyond the Academy, the group
formed an officially registered independent organization called the Baikal
Environmental Wave in 1992. Baikal Environmental Wave was among the first
environmental organizations to register as an official NGO in the region following the Soviet collapse in 1991.

From its earliest years, Baikal Environmental Wave had a strong transnational component, and this proved to be an important draw for locals into the
organization. The first members were recruited from an English language club,
and the ability to interact with Jennie Sutton and other foreigners, who regularly came to volunteer or provide assistance, continued to provide a strong
magnet for involvement. Curious locals, especially students, soon followed,
eager for the opportunity to meet people from abroad, which was a rare occurrence for Siberians at the time.

NASTYA: We had a department of foreign languages where Jennie Sutton,
our star, was. And she is the reason for our interest in the movement . . . So I came to Baikal Wave through the English language. It
turns out that I came to [environmentalism] not through biology, but
through English, and Jennie, and everything around that.

MASHA: I was studying at the Institute for Foreign Languages and a
family friend, a biology schoolteacher (and so knew the Wave), told
me that there is this interesting organization and there are foreigners
there with whom I could socialize.

SASHA: [I was taking an exam in English and was sent to Jennie.] She
would approve the quality of my translations; we had to give her a minimum of one thousand translations. She gave me environmentalist

texts, so I got into it that way . . . And I saw that there was a group that met with Jennie in her office to discuss them.

Although originally attracted to Jennie Sutton and Baikal Environmental Wave for the language practice, people stayed because of what the organization taught them about the environmental crisis, and they found themselves transformed in the process.

MASHA: When I only just started at the Wave, I sat, doing some translation, and they had a planning meeting. They were discussing an activity they were planning: to make rag bags to use in place of plastic ones. I sat and thought, "My God, what nonsense! What they are talking about is just nonsense!" . . . I really did love nature and I wanted to protect it, but like many of our people who love nature, I had no understanding that in your everyday life you can live in such a way (or not in such a way) so as to cause less harm. Those little things, like energy efficiency, where we get energy from and how it all works, how waste works. How it all begins with these little, simple, most basic things. I learned all this from the Wave. And in general, about social movements, about different political processes occurring in this country and in the world. I learned so much, and now, because of that, I think differently and perceive things differently.

For many of the Wave's early members, their attraction to experiencing international diversity became the door to a shared environmental collective identity.

As an official organization, the little band of Irkutsk environmentalists began their journey—learning what it meant to be an independent organization within a global environmentalist community. In a nation with virtually no modern history of independent organizing, they emerged ready to get to work.

JENNIE SUTTON: The Wave, right from the beginning, was a great learning experience. My feelings were acutely that this was all very exciting. *Perestroika* had been exciting. Then in the 1990s you have this collapse of everything around you, nothing in the shops, people losing their work, people going on the streets to sell things. One of our colleagues (who is now in Canada) was selling ice cream in the street because she wasn't being paid at her institute. She was a scientist, and she was selling ice creams in the street. . . . Society was collapsing around you, and yet we were *building* something! You felt you were

doing something positive. You were creating something in the midst
of this collapse. And I felt it was exciting, inspiring.

During this time, Western governments and private sector foundations were
eager to engage nascent nonprofit organizations in the former Soviet Union.
Many of my informants spoke of conferences, seminars, and trainings during
the 1990s, where Western organizations would provide opportunities to local
activists in "capacity building." The Soros Foundation, the Ford Foundation,
Initiative for Social Action and Renewal in Eurasia (ISAR), and the U.S. Agency
for International Development (USAID) were only a few of the most notable
bodies that were regularly cited by locals as cooperating in this endeavor. They
would sponsor workshops to instruct young nonprofits in Eurasia on how to
fundraise, how to work with the mass media, and how a nonprofit organization
might be organized. Baikal Watch was also an active presence, trying to build
capacity in the post-Soviet space.

> GARY COOK: [In Russia at this time] public or community activity of any
> sort without the government's permission or stamp of approval was
> very novel. There were no environmental groups. But there were envi-
> ronmentalists. Scientists. Teachers. But they were never able to get
> together to form even a nonprofit, nongovernmental group. . . . The
> idea was to help the movement start. . . . To help people understand
> what it means to be an activist, how to start activist groups, or even
> community groups, and to work with other sectors of society [such
> as] the government, the press, lawyers, scientists, anybody and every-
> body; to become a valid player.

What is important to note about this flurry of international support for organizational development is that Russia had virtually no history of independent
organizing. The Communist Party was the sole sponsor of any official activity
within the Soviet Union. When Clemens (1996) discusses activists as *brico-
leurs* in the creation of an organizational form, she assumes they are drawing
upon an extant "repertoire" (Tilly 1978) or "toolkit" (Swidler 1986) that Soviet
citizens simply did not possess. To that end, local Wave members were glad to
learn Western modes of organization, to acquire new "tools" for their "tool-
kits." However, they did not adopt organizational forms from the West whole-
sale. Instead, they took advantage of their liminal position to fully explore the
possibilities, carefully debating the implications of their choices.

In the 1990s, emerging from the heavy hand of the Communist Party,
Irkutsk environmentalists were at their most free to imagine and implement
organizational democracy—indeed, the Wave's resulting organizational

charter suggests that the shadow of Soviet authoritarianism was their most operative influence—a "repertoire" that they were keen to avoid. The Wave's organizational charter instituted the deeply democratic ideals of its membership. To prevent any one individual from dominating the Wave's activities and direction, the charter dictated that there would always be three co-directors at any given time. Also, for any decision requiring a vote, a quorum comprising two-thirds of the membership was needed to ensure that the taken decision would reflect the preferences of a majority. Prior to any vote, the membership would debate the issue at hand, and, while not always achieved, they strove for consensus.

> LYUBA: The Wave always had clear positions. They were, of course, born with great difficulty, since all decisions were made collectively. That is, they were all discussed until they came to some coherent outcome. We sometimes had seminars that lasted the whole week. We could discuss things from morning till night. My husband used to say— God, you will die from all of these conversations!

While the Wave membership had many opportunities to participate in trainings and seminars, learning Western-style NGO professionalization, they carefully weighed what they learned against their local knowledge to craft an organization that reflected their own needs and desires. With a nearly "blank slate," they both imported knowledge from abroad and produced new forms from their own imaginings. Not only was the result structurally democratic, it also enshrined education and advocacy as the means to social change. In retrospect, the focus on education was considered by some to be indicative of their naiveté: "We thought if the politicians just knew about these problems, they would do something about them," as one member said. But that naiveté is itself suggestive of the social empowerment former Soviet citizens felt in the wake of state and economic collapse. The 1990s were a terrible time for most Russians, but for those involved in the Wave, they had never felt so free. The organization grew and flourished, reaching is peak in 2002 when, according to its financial records, thirty-five people were actively employed in various environmental projects.

During the 1990s, while political institutions were in shambles and so-called New Russians fought their internecine battles for control of the various privatized economic assets of the old regime, members of Russian civil society worked to create an alternative social space at the local level. As individuals—and as organizations—they were largely unrestrained by opposing powers. However, the field of power is not always confrontational—it may be cooperative as well. Baikal environmentalists were instead constrained by their inability to make demands upon an incompetent and bankrupt state or a dysfunctional and disorganized industrial system. Despite the fact that civil society organizations thrived during this time, the environment they were striving to save saw little direct improvement.[1]

Nevertheless, once the political and economic powers recovered, civil society had to contend with these powers in new ways, none too cooperative. As the country stabilized, the strength that had accrued to Baikal Environmental Wave by virtue of its uniqueness and its cosmopolitanism[2] came into conflict with the growing might of economic and state elites. When this happened, it kicked off the first of a long series of attacks and attempted suppression of the organization by oligarchs and the Kremlin.

In 2002, the privately owned oil giant Yukos was attempting to construct a pipeline to bring oil from Angarsk to the east along the southern edge of Baikal. At the time, Yukos was owned by the oligarch Mikhail Khodorkovsky,[3] who went on to become the most notable of Putin's political prisoners, and a cause célèbre for international civil society. But in 2002, Khodorkovsky was the richest man in Russia and detested by domestic civil society for using his outsized power and influence in an era of rampant corruption. He acquired Yukos through Boris Yeltsin's notorious 1996 loans-for-shares program, and in 2002 he attempted to lay a pipeline from Angarsk to Buryatia through the Baikal watershed.

It was not the first pipeline proposed in the region, nor would it be the last, and in this, as in every other instance, Baikal Environmental Wave stepped into the fray to try to prevent its construction. This protest was the beginning of what has since become a hostile, adversarial relationship between the Wave and Russia's economic and political elites. Sutton was still a co-leader of the group, and she vocally opposed the pipeline. But her position as a British national gave fuel to the fire of pipeline proponents.

> JENNIE SUTTON: The media in Irkutsk and Angarsk were totally bought by Khodorkovsky. It was just awful. And they managed to latch on to *me*, and were writing that a *foreigner* was trying to use environmentalism to prevent the region from developing economically. I realized then that I could be a liability to the Wave, and I intentionally stepped out of a leadership position.[4]

As the nation's industrial sector recovered from the collapse of the 1990s, it once again emerged in the field of power to counter the push by environmental civil society against economic activity deemed damaging to Lake Baikal. Despite the evident allure that Baikal Environmental Wave garnered by its cosmopolitanism, which helped it when recruiting local support, its transnational character became suspect when pitted against economic recovery and growth. The force that allowed the Wave to develop to the heights it achieved by the late 1990s proved less forceful than an oil tycoon with deep pockets and a profitable agenda.[5]

To sum up, in the liminal space of political and economic collapse, civil society had the unique opportunity to invent itself, creating an organizational form that was radically democratic, expressing deep concern with concentrated power. Local activists learned from transnational actors and borrowed from their knowledge base, but at the same time were constantly discussing and debating each organizational choice. Activists in the 1990s were constrained mostly by their own recollection of the Soviet past and their desire to avoid repeating it; in their ability to imagine and implement an organizational form as a "frame" (Clemens 1996), they were at their most free.

Moreover, the beginning of a permanent, organized, environmentally focused civil society in Irkutsk was indelibly influenced by its access to the global plane. Through Jennie Sutton, an English-speaking foreigner engaging a scientific community near Baikal, the global discussion of sustainability intersected with local concern for the lake. The original members of Baikal Environmental Wave were initially brought together not because of environmentalism but because they wanted to learn English from a native speaker. And later, Sutton and other foreign volunteers continued to draw interest and involvement from local people for whom an exposure to cosmopolitanism was still rare. Foreignness was the first allure; it was only through discussing environmental problems and learning about sustainability that committed activists were made.

Still, these foreign connections threatened to hinder the organization. Sutton felt she could not continue in a leadership role out of fear that her "foreignness" could be framed negatively in the media to discredit the organization. However, these problems did not plague the organization until the Russian state and economy began to rebound in the 2000s, and Russian oligarchs such as Khodorkovsky began to set their sights on new development projects. When economic and political power were unstable, transnational connectivity was a source of organizational strength; when that strength was sufficient to threaten rising political and economic opponents, it became a liability.

Origin of the Tahoe-Baikal Institute

The Tahoe-Baikal Institute began in 1989, largely thanks to the efforts of a single individual: Michael Killigrew. Killigrew was a passionate and energetic peace activist and environmentalist who was eager not only to dream solutions to the world's problems but also to make them into reality. He was involved in a youth conference organized by Direct Connection US-USSR that aimed to bring peace to the Cold-Warring superpowers by building relationships between young people in these two countries. The conference took place in Helsinki in 1988; participants were asked to come up with ideas to promote peace, and one idea was a summer exchange for the purpose of scientific study and environmental protection.

Upon returning from the conference, the Soviet youth were able to win an audience with Mikhail Gorbachev, who supported the idea for an environmental youth exchange program. Ronald Reagan also stated his approval for the project, as did Perez de Cuellar, then secretary general of the United Nations.

With these powerful endorsements, Killigrew proposed the program to George Deukmejian, who was governor of California at the time.[6] Deukmejian, in turn, recruited Bud Sheble, who headed the California Conservation Corps, and Gordon Van Vleck, who ran the California Resources Agency. The team in California considered Lake Tahoe to be their own version of Lake Baikal: a crystal-clear mountain lake that residents saw as a point of regional pride, but likewise in need of protection. Between themselves, the three men pulled together a board of directors and a mission statement, and the Tahoe-Baikal Institute came into existence. Sheble served as the first chairman of the board and the organization was run out of his home office in Belmont, California, overlooking the San Francisco Bay.

Killigrew, Sheble, and the board of directors were prepared to develop an environmental education program at Lake Tahoe. However, for the program to be a cross-national exchange, one that would foster learning and mutual understanding on both continents, there had to be Russians willing and able to build a similar summer program near Lake Baikal. Killigrew used his contacts in the Soviet Union to seek out like-minded individuals in Irkutsk, and a delegation from the United States traveled to Irkutsk to draft a memorandum of understanding and to plan the first summer exchange.

An eager group of collaborators, collected from various branches of local government and scientific institutes, met the American delegation at the Irkutsk airport.

BUD SHEBLE: And from that point on, the vodka started flowing. Every meal, every meeting we had, breakfast, lunch, dinner, they brought out the vodka. I like gin, but I never liked vodka. But I drank it so much over there that I started liking it! [laughs] We'd be meeting, and then someone would stand up and say, "My dear friends..." and then we'd all have to stand and drink, and then we would go on with whatever we were doing.

The Russians quickly endeared themselves to the Americans who visited them. The society and culture that the California contingent encountered in Irkutsk was truly different, and very foreign, but the good nature, kindness, and hospitality that their hosts displayed moved them deeply. Sheble recounted one occasion when the delegation was visiting a small lake in the Baikal region, looking for a site for the summer work camp.

> BUD SHEBLE: They took us out to a little lake, just the tiniest thing, you could walk around it, where they thought they could make a camp for us ... They took us there, the prettiest little place. There were four or five Soviet gentlemen and an interpreter. I asked the question, "What's the name of this lake?" And they all looked at each other. Then they gathered together and you could hear them mumbling. Finally they stopped, and turned to us, and one of them said, "Tahoe!" It was absolutely hilarious.

The Russians' hospitality remained legendary to this group of Americans, even to their very last moments in Irkutsk.

> BUD SHEBLE: At the conclusion of our meeting in the Soviet Union, we said goodbye to everybody at a breakfast, drinking their vodka, and they loaded us into a van to go the airport. And we took off. We got just outside of town and we turned a corner on this road—and there they were again, standing on the road, waving and stopping us! We pulled over and they broke out the bottles again, and we all had a drink again. And I was told at the time that was a tradition, that they did that with guests. That they would run and meet them as they were going out of town, too! So as you can see we were becoming very fond of these people.

With the hard work of volunteers in Russia and the United States, the first Summer Environmental Exchange (SEE) program took place in 1991. Fifteen young people, half Russian and half American,[7] came together for a summer of environmental scientific study at Lake Tahoe, and then at Lake Baikal. The logistics were difficult to arrange, and at times there was worry that the visas for the Russian students to come to the United States would not be processed in time. In the end, all the pieces came together and the program was considered to be a great success. While the object of the project was environmental and educational, the overarching interest in the program, coming out of nearly a half-century of Cold War, was the sheer novelty of bringing together young people who had been taught to think of each other as enemies, and to see how the other lived firsthand.

BUD SHEBLE: It was an incredible experience because here were these Soviet kids, and we all feared the Soviet Union, but you had Soviet kids and American kids who find out that they are the same! They cared about each other, they cared about us. It turned out to be a fabulous thing.

While caring for the environment, young people were taught to care for other people living half a world away.

It was widely acknowledged that, when TBI got started in 1990, the Russians that had been recruited to the organization were ignorant as to what a nonprofit was, how a board of directors worked, or what it did. But they had enthusiasm for the idea and a strong desire for the program to succeed. Those involved spoke with evident pride of their work in building TBI-Russia.

NINA RZHEBKO: We tried to raise the level of the program every year. To that end, we tried every year to do projects that were a little bit more difficult and a little bit more interesting . . . We tried to do projects that were geared toward the future not just the present. We did lots of types of projects. And we tried to do these various projects in different territories. So there were ornithological projects, soil science projects, hydrological projects, all separate. There were projects in Buryatia, and some in the Irkutsk region, in different counties [v raznikh raionakh]. Sometimes we weren't even on the shore of Lake Baikal. We took a group to the wild regions of the Sayan Mountains, for example . . . So we had a sufficiently wide program.

From 1995 to 2005, the Russian side of the Tahoe-Baikal Institute was run by Nina Rzhebko with the support of a board of directors that was made up of scientists, government officials, and alumni of the program. All of them gave their support and effort voluntarily, just as the board in Tahoe did.

But the SEE program of 2005 had a number of problems that proved to be a tipping point in the relationship between the US and Russian branches of TBI. There were conflicts between the American students and the Russian staff. Rzhebko admits that she may have outgrown youth culture and could no longer relate to the students in her charge. "I no longer understood their jokes, their humor, their interests. And that is serious," she said. "When you have already graduated, when you hold a high position of authority, when you have your own children, and then to see that kind of humor, dropping one's pants and showing one's bottom, I simply . . . I saw that, as time went on, I started to look down on that [eto razdrazhat]." When the students asked for a day off, she would scold them, telling them that they were not on vacation. When they wanted to buy and cook their own food rather than what had been prepared, she was offended.

NINA RZHEBKO: So we wrote a sufficiently harsh letter to the directorate of Tahoe-Baikal. I think from our perspective it was righteously harsh. [But] Americans don't like it when you speak to them harshly, and that was something we came to understand. [*Amerikantsy ne lyubyat kogda s nimi rezko obrashchayutsya, a my tak i ponyali*]

The board of directors in the United States, who raised the funds for the Russian coordinators, announced that, rather than simply renew their contract with Rzhebko and her assistant, they would re-open the job as a competitive position. Rzhebko could give an application, but the board would be considering others for the job. Rzhebko declined to apply, as did her assistant.[8] In the end, another program alumna in Russia was hired by the American board of directors, and she had a vision for the program that was more in agreement with their own. This individual was presented to the Russian board as the new coordinator without their input or consent.

For the Russian board of directors, the insult was multifaceted and the injury cut too deep to be easily repaired. First, Rzhebko was beloved and respected by TBI community in Irkutsk. The Russian board felt that, especially after all her years of service to the organization, she had been poorly treated to be dismissed so suddenly and casually. Second, there was the added insult that they, as TBI-Russia, an independent nonprofit organization, had not even been consulted in the decision to dismiss Rzhebko, nor in the hiring of the new coordinator.

DIMA: You can ruin things in trying to improve them. There was a decision made in total that, in my opinion, was incorrectly taken. If you are talking about a private company, where the leader works for a paycheck, then you can take that person and fire them. But if you are talking about an organization where people are working fundamentally from their own good will, to take those people and say, thanks, see ya. Then those people will say, to hell with you, too! So, without any consultation, [for them] to say thank you but we don't need you anymore, the whole board left.

From the Russians' perspective, TBI-Russia had been an independent organization with its own vision and its own ambition. Its members were proud of what they had built. When Rzhebko was let go and a new coordinator was hired, they felt that their vision had been repudiated by the American side of the organization, and the American vision had been unilaterally imposed upon them. In response, the group chose to disband and to no longer volunteer their time and effort in support of TBI. While the organization remains regis- tered within Russia as a nonprofit, it ceased to exist de facto after 2005.

From the perspective of the American board of directors, it was not simply a matter of whether the students were scolded for dropping their drawers in jest. The board had long been concerned about the wide dispersal of students around Baikal. There had been some instances of individual safety and security far afield that worried the American directorate. They knew that they would be held responsible for any problems that took place within Russia. Having liabil- ity without accountability worried them. They felt that they could no longer risk ceding control of the program at America's borders. For them, the dissolu- tion of the Russian board of directors was unfortunate, but necessary. In their minds, the original vision of having two separate but cooperating national organizations was inherently flawed, and it was something that needed to be corrected.

The question of concern here is not which vision for the SEE program was superior, that of the American or Russian boards. Instead, it is the differen- tial power relationship between the two partnering organizations. The criti- cal difference that mattered in the eventual outcome was not in the power of each organization but rather each organization's position in the global field of power. Financial power is a generalizable power, but it is concentrated geo- graphically in the United States. State power is generalizable in a sovereign territory, but what that power can achieve geopolitically is not equal for all sovereigns, and the United States in 2005 had greater political leverage than did the Russian Federation. The field of power may be abstract, but its play- ers cluster in real geographic locations. Civil society organizations, which are materially located in different countries, also become instantiated in particu- lar spatial arrangements within the field of power. To that end, two comparable "sister organizations," though equal in domestic capacity, were highly unequal in the conflict over their shared future.

TBI-Russia disbanded. But the board of directors and alumni who had been the chief organizers for TBI-Russia prior to 2005 did not simply vanish. When you follow their individual paths after the dissolution of TBI-Russia, it becomes apparent that these disparate persons have gone on to further the development of environmental science and environmental protection in the Baikal region on their own initiative. They have started new organizations and new research centers, such as the Baikal Institute, Eco-League, and the Tahoe-Baikal Institute Alumni organization. They remain firmly committed to preserving the Baikal watershed, and now, thanks to them, there are more opportunities in the local community to work toward this end. They have their own students, their own colleagues and peer networks, and many attribute their ability in crafting such initiatives to the skills they learned at TBI.

> DIMA: I am very grateful for the experience that I gained there. The experience of organizational management and international program participation. Sure, it wasn't colossal, but still significant. And the experience of organizing projects, of project management, all that has played a major role in my future career. I now run a nongovernmental science center, which I founded. I am a professor, and I run a scientific laboratory. I have a ton of people there, and a ton of projects going on. And obviously, the roots of all that grow precisely out of Tahoe-Baikal.

Like that of Baikal Environmental Wave, the history of TBI shows the ways in which local civil society is shaped in relation to the global field of power. TBI was conceived jointly by Russian and American students. It was founded as a cooperative program between two separate nonprofit organizations—one in California and one in Irkutsk—working on two branches of a single summer scientific and environmental research exchange program. In structure, it suggested equality between members and participants in two national contexts. But their different physical locations materialized the unequal dispersion of generalizable power around the globe. The result was a kind of spatial inequality that subverted the founders' original intent. The US and Russian versions of TBI had different visions for the SEE program. But more importantly, they had differential capabilities for enacting that vision, located as they were in greater or lesser position in the global field of power. Russian activists could create the program that they desired on the ground, but they were relatively powerless in their ability to counteract the agenda of their international allies with a stronger field position.

However, despite what might be called the failure of the parallel sister-organization model of TBI, its implementation had a positive effect in fostering the general field of environmental activism in Irkutsk, and for the creation of Irkutsk as a space of transnational activism. Each summer for over twenty-five years, students from the United States and other countries would travel to Baikal for five weeks to become intimately associated with the lake, including its geology, geography, ecology, history, threats, fragilities, endurances, and beauty. Travel, tourism, and educational exchange breed place-based affective ties between dispersed individuals and distant points in foreign lands. Tourism is such an effective means to build transnational allegiance that foreign governments may intentionally sponsor tours to encourage these strategic affective ties (Kelner 2012). Through TBI, Baikal gained hundreds of "conscience constituents" (McCarthy and Zald 1977)—supporters world-wide who are ready, willing, and able to evangelize in their own countries on the lake's behalf.

Moreover, each year for a quarter century, a handful of local students from the Baikal region would gain the opportunity to travel abroad, and to experience environmental science and protection in the United States. They would build relationships and social networks with like-minded individuals, both in their own region and abroad, which would serve them throughout their continuing careers. In interviews, Russian participants of TBI repeatedly referenced a TBI colleague who later sponsored a research visa abroad or who wrote a letter of recommendation for an international grant. Participants from Irkutsk forged network ties that would alter the professional landscape of Baikal environmental science and advocacy. Also, those alumni who became involved in the planning and implementation of TBI repeatedly cited the experience as enormously influential in their ability to further the development of science, environmentalism, and transnational collaborative work independently and autonomously, with skills that continue to serve them well into the future. TBI alumni have spawned new organizations and environmental initiatives in Irkutsk and Buryatia. Although the joint project showed the negative repercussions of geographic power imbalance between partners in transnational activist collaboration, the participants in the transnational exchange program itself nevertheless acquired important experience, skill sets, and social capital that they then used to alter their domestic environment, as they worked to build a stronger civil society inside Russia.

Origin of the Great Baikal Trail

Although the Great Baikal Trail[9] organization was founded in 2002, the idea of building a long-distance hiking trail around Lake Baikal is much older. A number of Soviet writers, most notably Oleg Gusev and Valentin Bryansky, discussed the possibility of a long-distance trail that would circumnavigate the lake. Some intrepid outdoor enthusiasts during the Soviet era actually did hike, skate, ski, or boat around the lake without the help of a pre-existing trail. What was new in the creation of GBT was the notion of hiking trails as an environmentalist project geared toward localist, sustainable, economic development. GBT was created to organize and advocate for this environmentalist work.

GBT owes its origin to a small handful of individuals, and chief among them is Andrei Suknev. A native of Ulan-Ude, the capital of Buryatia, Suknev was an active youth in the late Soviet period, participating in what was then known as "do-it-yourself" tourism [*samodeyatel'nii turizm*]. These small clubs

of outdoor enthusiasts would train and equip themselves for wilderness sur-
vival. Suknev admits that even his own first reaction, when he heard of building a trail around Baikal, was that the idea was idiotic. "What would you need
to build? Geez, we're Soviet people, after all! We can hike without trails!"

But after the Soviet collapse, Suknev began to work as a tour guide. It was
only as he watched and participated in the development of tourism around
the lake that GBT seemed to him to answer the region's two most pressing
problems at once: economic development and environmental protection.
Low-impact ecotourism could help to sustain Baikal without sacrificing local
livelihoods. However, it would take a dramatic shift to turn what had been a
marginal do-it-yourself hobby club into the kind of tourist industry that would
bring some economic self-sufficiency to the villages around Baikal, while
remaining environmentally sustainable.

When he first began to proselytize the notion of building hiking trails,
Suknev said, he encountered resistance from locals who were unclear what
exactly he meant.

"They would say, 'A path?' [*marshrut?*]. And I would say, no, a path is a path.
A trail is *infrastructure*. It is specially done so that people won't hurt themselves, and won't hurt nature . . . It [represents] a better quality of life."

Suknev converted several of his friends to the idea, but they made no further
progress on the project until it found a catalyst in Gary Cook at Baikal Watch.

> GARY COOK: This one lunatic friend of mine came to me and said, "I want
> to build a trail all the way around Lake Baikal." And I said, "You're
> a lunatic!" But he said, "No, listen to me. We'd like to try this out."
> It was interesting because he and his colleagues theoretically under-
> stood what they wanted to do, but to get from the idea to actually
> building the trail was a quantum leap. So I felt that there has to be a
> very light velvet glove that [Baikal Watch] applies to this from the out-
> side. We were going to have to give them lots of support to understand
> what this needs. How do you build a trail? How do you design it? How
> do you work with local communities and local parks? [How] to get
> that trail built, to make sure it's the trail you actually want. And most
> importantly, how to build a trail when you have *very limited resources*.

Cook's philosophy toward American involvement in Russian environmentalism has always been that of assistance and support. Baikal Watch involves
itself only upon invitation from local Russians, and it always asks the locals
to take the lead. "How can we support you, so that you achieve what it is that
you want to do?" is the question that Cook poses to his Russian collaborators. He is cognizant that he is working in a foreign country, and he considers

Russian environmental problems and—to a degree—their solutions to belong to Russians themselves. He strives to avoid the potential for an imperialist approach to international aid and development, hoping instead to empower local activists to become leaders in their home communities.

Nevertheless, as his comment about the "light velvet glove" indicates, Cook does have a vision for what needs to be done to see projects succeed, and his vision does guide the trajectory of the endeavor through his capacity as a helper and supporter of Russian environmentalists. When it came to the creation of GBT and the shape it eventually took, Gary Cook played a central role. Suknev is rightly called the "father" of GBT, but Gary Cook is undoubtedly its "godfather."

His first action was to bring together powerful players in the Baikal region, to show them the vision, and to explain how it might be made a reality. The hope was to produce collective buy-in and to garner support for the project among power figures and key stakeholders in the region. With a grant from the Foundation for Russian-American Economic Cooperation, Cook and Suknev began to build their case through two cross-national exchange programs. First, in the fall of 2002, American experts working in national parks, the forest service, ecotourism, and trail building came to Irkutsk for two weeks to talk to their counterparts about the potential development of ecotourism in the Baikal region. A few months later, in January 2003, a similar delegation from the Baikal region came to the United States. The grant covered an all-expenses-paid educational trip up and down the West Coast, where the Russian representatives could see ecotourism in action and meet with groups like Earth Corps in Seattle and TBI at Lake Tahoe.

Toward the end of the visit, the Russian delegation was taken to a classroom in an old school near Lake Tahoe. Gary Cook instructed them that they were to talk among themselves and then decide whether they would try to create such a thing as the "Great Baikal Trail." The decision to proceed, and the responsibility for the project's implementation, would be theirs alone. Cook left the room, closing the door behind him. When the group emerged three hours later, it did so having planted the seed for GBT.

The next question was how these trails were going to be built. Based upon American experience in trail construction, Cook had a ready answer. "I said, 'You are going to have to introduce something very novel in Russia and that is the concept of volunteerism' [laughs]."

The Russian word for a volunteer is *dobrovolets*, literally meaning "someone who does something in a spirit of good will." However, in the Soviet Union, citizens were "volunteered" by the government to do work projects outside of their regularly paid hours to supplement the state's need for additional labor at particular intervals, for example, during harvest season. While the word translates literally as people motivated by their goodwill, it took on an ironic, negative connotation in the Soviet Union of begrudged work forced upon an unwilling populace. The dreaded *subbotniki* (Saturday volunteer work day) helped discredit volunteerism as a concept in Russia's post-Soviet "cultural toolkit" (Swidler 1986). There were individuals in Russia who were ready and willing to give their free efforts toward a good cause, but they lacked an obvious outlet for their energies.

As an American long accustomed to the power and possibility that can be harnessed in volunteerism, Gary Cook was confident that such a massive infrastructure project as GBT would never get off the ground without volunteer labor. The question, from his perspective, was how to overcome the cultural mismatch between Andrei's vision and the negative connotation associated with the *dobrovolets* in the Russian mind. To surmount this obstacle, Cook again turned to the enticement of cross-cultural exchange.

> GARY COOK: I said, "Listen, let's try to be imaginative. What if we brought a bunch of international volunteers [to Baikal] and advertised it as: Come see Siberia!" Even in 2000 there weren't a lot of Westerners who had been there . . . [Westerners] not only want to see Baikal and do something positive, they want the real Russian experience and what better way than to hang out in a camp with a bunch of Siberians?

Thus, foreign volunteers could compensate for the general absence of volunteerism in Russian culture as a whole. The foreign volunteers would not only pay their own way but would also cover the costs for Russian recruits. This scheme provided an incentive that enabled GBT to recruit Russians who might not otherwise volunteer to help on the project. Locals were told they would get a two-week, all-expenses-paid trip to Baikal and the opportunity to meet people from all over the world, as they worked side by side to build a hiking trail in the woods. In the end, they succeeded in recruiting locals—primarily students—to join the project, and the first summer trail-building program took place in 2003. To help overcome the negative connotation associated with the word *dobrovolets* and the forced free labor it implied, GBT referred to its helpers by the English word—calling them *volontyory*.

Although the closed-door, Russians-only discussion that took place in the school outside Tahoe had come out in favor of creating the GBT project, the Russians' commitment to this endeavor was still hesitant. Although convinced of the model's efficacy within the United States, most still viewed the possibility of its success on Russian soil with skepticism. But some participants were converted to the idea and embraced it with enthusiasm. Among them was Ariadna Reida, who would eventually serve as the organization's first executive director.

At the time of the Russian delegation, she was serving as the group's translator. She was one of the few who believed in its potential. Most of the others simply could not comprehend volunteerism as a concept.

> ARIADNA REIDA: The people of power who got together [in Tahoe] said, okay, we want to give it a try next summer, the summer of 2003, to do international camps, to actually build trails. They said, "We will give it a try," but nobody believed that people would actually come and work for free.

With the willing, if skeptical, agreement of key stakeholders in the Baikal region, the first GBT work trip took place in the summer of 2003. To the surprise of the skeptics, the program succeeded. Volunteers showed up, and trails were constructed. Certainly there were problems and difficulties, but the response from all involved was positive. In fact, according to Reida, the volunteers' chief complaint was that they were not able to work *more*. The paid employees in the protected territories around Baikal were so doubtful that volunteers would actually come, they did not provide sufficient tools for the group. When tools were available, staff tended to monopolize them and would not share them with volunteers, because they were "tourists." But trail building was accomplished, and those who had been skeptical were falling in line behind Andrei's vision, acknowledging its potential.

At the end of the summer, the group met again to assess the summer program, and Reida again served as a translator and group facilitator.

> ARIADNA REIDA: There was a lot of interest and a lot of excitement about the project. It was really funny because locals were saying, "They actually came! These crazy foreigners, they actually paid to work!" . . . So there was some publicity, and the excitement that people came. And they came not only for the two weeks [of the work camp], they would stay and go as tourists somewhere else on the lake. So [people could see that] it was building the economy.

The group decided to continue the projects into the future, making improvements in planning and execution so that the next summer would be even better.

Young Russian students who participated in 2003 kept the spirit of camaraderie and volunteerism alive by forming a club. They began holding weekly meetings to share stories, reminisce, drink tea, eat cookies, laugh, and eventually develop new ideas and projects. This pool of steady volunteer supporters changed the shape and spirit of GBT. Unlike the Wave, it would not be primarily a professional organization, relying on paid staff and grant funding. And unlike TBI, its activities would not be mostly confined to the summer months. The club became the springboard for a new collective of steady membership. Members would work year round, developing and conducting trainings for crew leaders and translators, planning summer projects, and getting involved in environmental education and clean-up efforts wherever such opportunities arose. The pool of *volontyory* continued to grow, largely by word of mouth. The years 2005–2007 saw a huge spike in club membership. Club meetings would see upward of forty people. Friends drew one another in, and once a part of the organization, friendship kept them there. Members would happily cite for me the marriages and the couplings that occurred through GBT. It is a thriving community of hard-working volunteers, eager to help others and be out in nature, quick to laugh and to have a good time—their eyes on the future with optimism.

GBT was long a dream of local residents around Lake Baikal, but the creation of the local organization and the form it took was fully intertwined with the global. Gary Cook was the catalyst that brought together the various parties whose support and cooperation were necessary for such a project to succeed. Key figures were lured by the free trip to the United States and, once there, they were shown repeatedly the success of an ecotourist model of sustainable development. Although Cook made it abundantly clear that neither he nor any organization in the United States would be playing a leadership role in GBT, he also made possible the conditions for important gatekeepers in Russia to publicly state their commitment to Suknev's project.

Still more importantly, it was Gary Cook who pushed for what has become a central aspect of GBT's organizational character: volunteerism. It is widely acknowledged by GBT members that the project could never have gotten off the ground without the presence of international volunteers. There was little doubt in anyone's mind that, as of 2003, Russians would not have volunteered in sufficient numbers simply for the joy of building a trail, and they certainly would not have paid money to do so. The money paid by foreign volunteers allowed Russian students to attend for free. Some of the organization's most committed and involved volunteers today first attended for free and would not likely have participated without that bonus. Some even reported that they went on two projects their first summer: there was not sufficient recruitment for all the free slots available to Russians, so after completing their first project, they could immediately join another. But every year, more and more Russians have been signing up—and there are even projects where Russians significantly outnumber the foreign volunteers, despite the fact that they are now required to pay their own way. Although critical at the outset, once established, the dependence on foreign participation was diminished.

Conclusion

The opening of the Soviet Union brought major changes to local environmental activism around Lake Baikal. Irkutsk was already home to a strong environmentalist community when the transition began. But the collapse created a new arrangement of players in the field of power. This meta-field becomes visible in the subsequent development of local Baikal activism. This chapter traces the field of power across two dimensions: the spatial and the temporal. Spatially, the Soviet collapse introduced local activism to the global plane, allowing domestic environmentalists to draw strength from access to this new level of social activity. Temporally, the field of power changed shape over time from the weak and unstable state of financial and political powers in the 1990s to their resurgence relative to civil society in the Putin era.

The opening of Russia to global interaction created many different opportunities for civil society to grow in strength. First and foremost, Baikal environmental organizations were quick to reach out to peers abroad for information, and these contacts made them appealing to locals who were eager to experience cosmopolitanism. Global peers provided resources that were in short supply domestically, whether those resources were financial (as during the 1990s) or cultural (as when international volunteers supplemented the volunteer spirit that Russian culture lacked). Baikal activists utilized the global plane in novel new ways to enact their local vision.

But the global is also a space that contains the field of power. In any transnational collaboration, individuals and organizations participating in shared projects cannot fully ignore the inequality inherent in the global power structure. TBI's American board members were not superior in commitment, scientific knowledge, or organizational skill to their counterparts in Irkutsk, but they possessed another type of power that accrued to them simply by virtue of their geographic location in a spatially instantiated global field of power. The United States, as the strongest economy in the world, provided the money needed to run the entire TBI operation. As the United States is also the most influential political force, American assumptions about the proper scope of program activity and legal liability are hegemonic. The endorsement of American donors and participants was necessary for the SEE program to continue, so much so that unilateral action by the Tahoe board of directors was seen as entirely justified, even if it dissolved the very existence of TBI-Russia. Transnational activist collaborators, even with the best intentions, must contend with the field of power and the geographic inequalities that it engenders in contemporary collaborative activism.

The field of power acts more directly, albeit more subtly, on the other two cases under study here: Baikal Environmental Wave and the Great Baikal Trail. When the Wave was founded, the political and economic structures within Russia were in shambles. Yet domestic civil society was buoyed by support from abroad. The very newness of democracy and independent organizing gave individuals a sense of personal empowerment that strengthened the claims they could make upon businesses and political leaders. The 1990s were a high point for civil power within the field of power. During this period, Baikal Environmental Wave was founded and flourished.

However, the organization has suffered in the years since the early 2000s, which is also the period associated with the re-emergence of economic and political power within Russia. Some of this weakness is the result of retaliation and suppression by the re-emergent financial and political powers. The fight over Khodorkovsky's pipeline and the Angarsk media accusations against the Wave are only one example of retaliation by economic elites. It would not be the last of the Wave's encounters with repressive and powerful opponents (see Chapter 7).

But there is also reason to suspect that the recovery of the economic and political forces in Russia altered the social desirability of strong advocacy organizations. During this same period when Baikal Environmental Wave began its descent from a high point in the late 1990s, we witness the emergence of the Great Baikal Trail, which proceeds to flourish. The mission of GBT is economic development through ecotourism, and as Suknev explains above, it was the direct result of his own observations of the apparent conflict between preserving Baikal and developing economically. As the country began to establish itself in the global economy, including the international tourist industry, Baikal was facing new pressures. GBT works synergistically with economic forces—making accommodations rather than demands. This style of activism is more popular and attracts more support in the Russia of the 2000s.

Not only is the model of volunteerism and economic accommodation more popular with power elites in the political and economic spheres (Hemment 2015; see also Chapters 5–7), but it also resonates more with the population, particularly youth. Most of the Wave's current members are of an older generation. Their younger members today joined through their prior involvement in radical politics. Already on the social fringe, they are willing to take on more confrontational advocacy work. But young people make up the majority of GBT. I asked one member of GBT why she did not volunteer regularly with the Wave. She replied:

KATYA: The Wave has a position, and GBT . . . well, GBT has a family. We have all kinds of things that we do, and we don't necessarily have a stand on them. One day we are helping kids get office supplies, the next we are building trails, and then someone asks us to pick up litter and we do that, or to organize an event at a fair. We just help each other out. And the Wave needs to have a position on whatever they do;

they know what it is they are against. You know, *that* is what it comes down to: the Wave is "against," and GBT is "for."

But the tables were turned only fifteen years prior, when the Wave was a strong and flourishing organization, while still taking their advocacy with utmost seriousness. The flexibility of GBT's organizational model, which is perceived as positive and proactive, comes at the expense of stricter demands on the behalf of nature. The same GBT volunteer quoted above went on to commend the Wave as the better *environmental* organization, even if it is, in her view, less pleasant to participate in.

KATYA: They practice what they preach . . . They have a motion sensor to turn water on and off in the sink, and they are good about turning off the lights and unplugging computers. At the GBT office, I feel a little hypocritical, because it is an environmental organization, but the office is not environmental at all. There is even this big crack right in the front window. Have you ever noticed it? Here we are supposed to be handing out brochures on saving energy, and we ourselves have a big crack in the window going outside.

But GBT's apolitical and accommodational character is decidedly reminiscent of Soviet-style environmental protection activities (see Chapter 2) Planting trees, cleaning litter, holding events at a fair—these were the quintessential pursuits of the *druzhiny*[10] and VOOP. They are feel-good activities, and certainly they are very helpful. But they lack the critical element that the Wave has possessed since its early years. The success of GBT in the 2000s attests to the changed balance within the field of power in Putin's Russia; accommodation to the current powers is preferred over assertive alternatives. The Communist Party allowed only certain types of environmentalism in the Soviet Union: those which were considered non-threatening to their power and agenda. It is telling that the organizational form that is ascendant today bears a strong resemblance to the student environmentalists of the Soviet era, while organizations emphasizing advocacy and confrontation seem increasingly untenable. Domestically, the field of power is again constricting, away from an empowered civil society.

Environmental activists do make organizations but, to paraphrase Marx's oft-repeated aphorism,[11] they do not make them just as they please. Civil society and the social organizations that it builds are powerful only in relation to other generalizable powers in the field. When the economy and the state are weakened, civil society is at its most free to pursue its own agenda. It can stake a position that makes demands upon other powers to produce change by means of strong advocacy organizations. When economic powers are ascendant, civil society may be more accommodational, emphasizing non-threatening, cooperative projects. Thus, the field of power conditions local civil society through the relative waxing and waning of the strength of its power source and its geographic location in a spatially differentiated global field.

4

A Tale of Two Lakes

Among the goals of transnational activism is the sharing of "best practices," which are methods or templates for action that have repeatedly proven successful (cf. European Commission 2014). The hope is that, in sharing best practices, weaker, younger, or less experienced organizations can avoid the pitfalls of trial and error, and in so doing, more quickly establish themselves and achieve their ends. But good intentions do not necessarily produce good results. Scholarly investigation into the sharing of best practices has yielded important criticisms. At the extreme, the best practice paradigm has been accused of foisting particular structural conditions on the Global South that are conducive to Western neo-imperialism (Goldman 2001, 2007). But more often, best practices are faulted for ignoring local conditions and imposing cookie-cutter answers to contextual problems that often require more subtlety in their solution (e.g., Wareham and Gerrits 1999).

This chapter examines an attempt at the sharing of best practices between communities around Lake Tahoe and Lake Baikal. While these two lakes have a historical relationship and have been designated "sisters," their differences far exceed their similarities. Replicating Tahoe's success at Baikal may be quixotic, given their unique and divergent contexts. Such an interpretation of events supports the critical literature on transnational NGO collaboration. However, the attempt to share "best practices" transnationally has a major, if frequently unacknowledged, *unintentional* effect, beyond activists' stated goals. Namely, transnational interaction has the opportunity to change how people think. Cooperation across borders may help participants to overcome entrenched fatalism and increase faith in the power of ordinary people to promote social change.

No particular method or skillset was actively diffused through the interchange between Tahoe and Baikal. What did change was how local residents around Baikal thought about their own circumstances and efficacy. Transnational collaboration helped to create the "cosmopolitan vision" (Beck 2006) in the most remote of places. Russian villagers, who prided themselves on skepticism, and whose social imagination ran up against the barriers of the tried and true, experienced a small mental shift when brought into contact with activists abroad. By their exposure to a different culture and to a milieu that respected progressive thought, Russian villagers expanded their sense of what was possible. In a country with weak civil society (Howard 2003) and a highly skeptical public (Mishler and Rose 2005; Goodwin and Allen 2006), transnational exchange created the space to imagine alternatives. Civic engagement, which had previously been eschewed, was brought into the realm of possibility.

There are limitations on the extent to which such a mental shift may take place. It does not automatically arise in every transnational interaction, nor is it necessarily sustained beyond the space of the interactive process. Local conditions necessarily influence the interpretation of the transnational experience. But the possibility has far-reaching implications. Not only is such an effect important in authoritarian regimes where life experience teaches the public to surrender to the will of the state, but it also could have critical consequences for the larger field of power in the era of globalization and authoritarianism. Since the general public's perceived possibility informs their willingness to engage in voluntary action, transnational collaboration has the *potential* to augment civil society's generalizable power.

A Tale of Two Lakes

Lake Tahoe and Lake Baikal sit on opposite sides of the world. And yet, as Chapter 3 shows, there is an invisible tie that binds them, made manifest by the annual migration of young people from one lake to the other through TBI.[1] The connection between Tahoe and Baikal was originally contrived by the founders of TBI: Baikal was a national priority for Russian environmentalists, and, for the purposes of building a cross-cultural cooperative environmental endeavor, their American counterparts in California thought of Tahoe as a lake in their home state that was also both beloved and in need of environmental preservation and rehabilitation.

Once the decision was made to link Tahoe and Baikal, a number of similarities became apparent to justify the selection. First, both lakes were formed through tectonic processes. They are both located in mountainous terrain that hosts many similar flora and fauna, such as coniferous forests, bears, moose, and foxes. Indeed, many TBI participants told me that the lakes are so similar in appearance that they find it difficult to tell which lake is which in their own photographs. Each lake is located on the border of two different governing authorities: Tahoe is split between California and Nevada, while Baikal lies between Irkutsk and Buryatia. Both lakes are prized for their clarity. Both lakes have seen mining and logging activities in their watersheds, although for Tahoe this threat is now only historical. And both lakes are host to recreational tourism.

At this point the similarities end. Most obviously, Baikal dwarfs Tahoe in size, as it does most lakes on Earth. When comparing the volume of Tahoe to that of Baikal, TBI members hold a golf ball next to a beach ball. Tahoe is small enough in circumference that a car can circumnavigate it in a day, whereas a trek around Baikal is not only impossible by car (because there are no roads around most of it), it would also take several days to travel the eight hundred miles mapping its circumference. The difference in size gives rise to different environmental hazards. Roads and development have made erosion the chief concern at Tahoe, with silt destroying water clarity and upsetting ecological balance in the relatively shallow body of water. Erosion of Baikal's islands is a problem for its residents (human and nonhuman), but silt has little impact on the mile-deep waters and the lake's profound clarity. Baikal, on the other hand, has been forced to face the hazards of industrial and agricultural effluvia, which Tahoe has never faced. While Tahoe is host to some endemic species, it is not the "Galapagos" that Baikal is, with endemics by the thousand.

In addition to these ecological distinctions, there are still more profound differences in the human societies that surround each body of water. Although both lakes are dotted with cities and towns, the level of development one can expect to find in each location varies dramatically. While there are locals who make Tahoe a permanent home, the lake is better known as a rustic retreat for peripatetic, jet-setting urban elites. Less than an hour from Reno on the Nevada side, and three hours from San Francisco with its major international airport, Tahoe is embedded in a rich network of First-World recreation. Paved and well-maintained highways connect the towns around Lake Tahoe to cosmopolitan capitals, and a steady stream of traffic in both directions renders Tahoe, for all its rustic charm, a place of urbanity. As a vacation home hub for Bay Area residents, real estate around the lake can range from a median of $150,000 in South Lake Tahoe to a median near $630,000 in Incline Village in the north (Carey 2013). The latter also has the notorious distinction of being a tax haven for shell companies and for wealthy Californians, who claim this Nevada address as a permanent residence (Liu 2008). Even in South Lake Tahoe, where the working class congregates and which has a poverty rate close to 20 percent, one can expect a basic level of development, which has become standard in the United States: plumbing, running water, stable electricity connections, Internet hookups, central heat and air, gas stations, convenience stores, and grocery stores.

In contrast, the villages and cities surrounding Baikal lack Tahoe's connectedness. There is no quick access from Moscow or St. Petersburg to the shores of Baikal. Most of the population centers along the lake are villages that seem to have changed little across the decades. Many lack plumbing. In the summer months, electricity is only sporadically available. Accessing villages on Baikal often involves long, bumpy rides in a crowded minibus [*marshrutka*] over dusty roads according to a schedule that can run as infrequently as once or twice a week. Some villages are only accessible by boat. There are two cities on the shore of the lake (Baikalsk and Severobaikalsk), but travel between them takes two full days by train. To live on the shores of Baikal is to be not merely rustic but remote.

Not only do the regions vary considerably in their levels of economic development, they also exist in very different political cultures. Americans have a long history of democracy, civic engagement, and responsive government (Tocqueville 1981). Russians, on the other hand, hold onto certain legacies of Communism and Empire, which range from bureaucracy to corruption. Russians generally have ambivalent feelings about democracy and little faith in their government (e.g., Mishler and Rose 2005; Goodwin and Allen 2006). More often, the state is something that is to be bypassed and avoided. Given their profound economic and political differences, the human ecology of Tahoe and Baikal create two very different and distinct "life-worlds" (Habermas 1981). That they could share common strategies for addressing social problems is perhaps surprising, yet that is just what two local NGOs at each lake aimed to achieve in a joint collaborative project.

A Project for Mutual Assistance

In 2012, the Wave and TBI received funds from the Eurasia Foundation for a year-long project that would link together three communities around Lake Baikal with organizations in South Lake Tahoe for the sharing of best practices and cooperative development. The Wave and TBI planned to reinvigorate the historic relationship between Tahoe and Baikal while conducting research to uncover which local resources could lead to environmentally sustainable development in each of these communities.

The stated purpose of the exchange was for communities that have faced a similar situation—the need to develop sustainably due to their proximity to a protected lake—to share best practices. The grant application states that the project's goal is "to form mechanisms for international mutual assistance for local communities, facing complex socio-ecological and economic conditions, through a vision of sustainable development in communities based in preserving the eco-systemic condition of two lakes—Lake Baikal in Russia, and Lake Tahoe in the United States." The language of the grant suggests cooperation and the exchange of information and expertise.

In this endeavor, TBI and the Wave were enacting a strategy for development that has become paradigmatic among the nonprofit community in the West and among transnational activist networks. As problems have become globalized, so too have attempts to solve them (Smith 2008). There is a strong desire to find what works and to share that knowledge with others in the hope that they can replicate past success (Seidman 2007). The evidence of transnational sharing of practices can be found in the "isomorphism" (DiMaggio and Powell 1983) that replicates similar norms across the globe, in policy and practice (cf. Dobbin, Simmons, and Garrett 2007). Scholars have shown that NGOs are as responsible as global business or national governments for the growing isomorphism (Schofer 2003; Schofer and Hironaka 2005; Schofer and Meyer 2005). As norms become isomorphic, they then can change the practical strategies of those people whom NGOs aim to serve, furthering the hegemony that Western organizations them-selves began (Elyachar 2006). Activists have encouraged the spread of certain "best practices" for labor and environmental standards among businesses world-wide, through a wide array of strategies and tactics (Blair and Palpacuer 2012).

However, there is a growing concern among scholars that what is com-monly deemed a "best practice" may fail in ways that are culturally specific, or that they may be even transnationally coercive. AIDS advocacy organiza-tions in Ghana that attempted to follow the successful media campaigns of Western nations could not convey their message in a built urban environ-ment that differed from that of the United States and Europe (McDonnell 2010). Attempts to provide clean drinking water in the Global South effec-tively restructure national governments and local institutions toward open-ness to Western corporations (Goldman 2001, 2007). Whether the problem is Western neo-imperialism or simply contextual ignorance of local processes, there is a growing concern that "sharing best practices" may not be a particu-larly useful practice in itself.

This chapter evaluates an attempt at transnational development work in accordance with the goal of mutual assistance—and shows the shortcom-ings of sharing strategies irrespective of contextual differences. However, the attempt to unite communities of difference for mutual assistance has an argu-ably beneficial, yet wholly unintentional, side effect that is potentially more profound than the sharing of best practices itself might have been. Despite the failure to transfer a "best practice" from one place to another, the very practice of sharing can potentially open minds to new possibilities and new terrains of imagination. Certainly, this outcome is not guaranteed. There are potential pitfalls to the self-same strategy: the interaction may result in disappointment and disillusionment. And local conditions may work against the effect pro-duced. But in studies of transnational encounters, the opportunity that collab-oration provides to reimagine one's own circumstances is often ignored. The expanded horizon it provides may be a worthy end in its own right, beyond any particular "best practice" conveyed.

The Wave-TBI project for mutual assistance was primarily devised and developed by Marina Rikhvanova at Baikal Environmental Wave, and it was clear that her own hope in the cross-national component was to encourage Baikalsk and the villages of Goloustnoye (Boshoe and Maloye[2]) to emulate South Lake Tahoe and its success in fostering sustainable geo-tourism. When I first discussed the project with her, as it was just getting started, Rikhvanova explained it this way:

> MARINA RIKHVANOVA: People in Baikalsk are very concerned about the paper mill closing because it has been the center of their economy. But Bolshoye and Maloye Goloustnoye had a similar situation. Most people worked for the *leskhoz* [state forestry company] that operated in that region. But when the national park formed, it became illegal to harvest wood in the territory of the park. The *leskhoz* had to scale back and many people were laid off. People in these towns generally think that a good job is a job in a big factory, like the *leskhoz* or the paper mill. They often don't see the resources and the potential that they have to develop economically in their region without some big industry. But in Tahoe, they have an economy built around tourism. They have managed to protect their lake and their forests, and they still develop economically.

For Rikhvanova, Tahoe is a success story. Her aim in connecting the towns around Lake Baikal to groups in Tahoe was education and emulation. Tahoe had things to teach Baikalsk and the Goloustnoye villages about the economics of sustainable tourism. Over the decades, Lake Tahoe had recovered from environmental degradation while maintaining economic viability. Residents living near Baikal needed to hear that such an outcome was possible and learn how it could be done.

But before they could connect these Russian sites to Tahoe to discuss mutual problems, they needed to find out what those problems were. First, the Wave needed to engage the public on its own. The environmental organization in Irkutsk sought to mobilize villagers of Bolshoye Goloustnoye to protect themselves and the environment from external efforts at development. The Wave conducted a series of field excursions to Bolshoye and Maloye Goloustnoye to encourage involvement, but their efforts repeatedly ran ashore in the face of village intransigence and a general pessimism about the possibility of change. Despite the Wave's repeated efforts, the villagers remained solidly cynical and fundamentally fatalistic, generally unwilling to get involved.

Cultivated Disinterest in Bolshoye Goloustnoye

There was a reason that the Wave was so interested in this one small village of six hundred people. The Irkutsk *oblast* [region] was in the process of setting up two special economic zones (SEZ) on the shore of Baikal. The SEZ would be an area where developers and investors could receive tax breaks and other incentives for establishing tourism and recreation enterprises in the region. Bolshoye Goloustnoye was one suggested site for the SEZ, and the Wave wanted to make sure that villagers' interests were represented in the process. They generally feared that SEZ-sponsored tourism would be environmentally harmful rather than sustainable, and they hoped that Bolshoye Goloustnoye villagers would work with them to ensure that outside developers would not run roughshod over Baikal and its indigenous people. Villagers were aware of the planned SEZ. They were told that it would bring running water and a sewer system to the village, which sounded quite nice from their point of view. Mostly, they were irritated that the land that was slated for development— which, when "seeing like a state" (Scott 1998), looked like empty space—was actually prime grazing land for the village cattle. The SEZ would essentially be enclosing the commons.[3] While activists and capitalists disputed the best means for economic growth, the people of Bolshoye Goloustnoye simply wanted to know where they could feed their livestock. The Wave looked with foreboding at big business's appropriation of local livelihood, and they wanted to assist the residents in finding a secure path forward in this brave new world.

In September 2012, the Wave made a research expedition to Bolshoye Goloustnoye, and I followed to observe the process. Bolshoye Goloustnoye is a village of about six hundred people on the shore of Lake Baikal, approximately two hours from the regional capital city, Irkutsk, by minibus along a bumpy, dusty, tortuous road. I traveled there with Artur, a thin, taciturn man with dark hair, a short beard, and piercing blue eyes. With us was Tanya, a sociologist that the Wave had contracted to perform the focus groups and interviews. The country road to the village of Bolshoye Goloustnoye was beautiful, with rocky cliffs rising up on either side. The road would often follow the small Goloustnoye River, and the autumn leaves that adorned the surrounding trees were golden.

The van lumbered along the winding road, lifting over one of the high hills and then descending into the river delta that opened out onto the shore of Lake Baikal. Nestled beside the delta, in a wide valley of steppe, sat the village of Bolshoye Goloustnoye (Figs. 4.1 and 4.2). The town was mostly made up of wooden, one-story houses, often with outhouses and small agricultural plots beside them, surrounded by wooden fences. Painted crenulations adorned some of the windows and eaves, in traditional Siberian fashion, but such sporadic

Figure 4.1 The village of Bolshoye Goloustnoye, population 600, on the shores of Lake Baikal. Photo Credit: Author

Figure 4.2 A very young resident of Bolshoye Goloustnoye. Photo Credit: Author

decoration did little to mitigate the general dilapidated ambience of a village that mostly survives on subsistence agriculture. Cattle and stray dogs roamed freely on the dirt roads, and children would ride along among them on bicycles, sometimes with another child sitting on the handlebars or perched on the back.

The Wave planned to conduct two focus groups and a community meeting over the course of several days in the village. The first focus group was to take place among students in the oldest grades at the local school (8th and 9th graders; those wishing to complete their high school diploma must travel to Irkutsk to do so). Since many villages in Russia currently face the problem of depopulation, the Wave hoped to learn from the students themselves what they liked and disliked about village life, and whether they planned to return to Bolshoye Goloustnoye after completing their education.

The schoolhouse had several classrooms, each imbued with a homey feel. The walls were wood-paneled, with floor-to-ceiling windows to allow in the natural light. Back walls were adorned in potted plants. Students—boys and girls—sat in their white and grey uniforms; most were alert, with straight backs and pleasant expressions, eager to please their out-of-town guests.

The focus group proceeded, and the students described the positive aspects of village life: being close to nature, on the shore of Baikal, with fresh air and homegrown food. Most still planned to leave the village, saying that there was nothing to do and nowhere to work. Among their chief complaints, one that was also repeated frequently by adults in interviews, was the lack of a preschool in the village. When students brought up the fact that Bolshoye Goloustnoye had no preschool, Tanya asked the students whether there was anything they themselves could do to fix the problem.

"No," several answered, shaking their heads.
"The administration [government] should do something about it," said a
   girl in a grey jumper, sitting in the front row.

When we left the school, it was obvious that Artur was in an irritable mood. The school housed the only computer in the village, which had a slow, dial-up Internet connection. Artur told the principal he planned to keep her informed as the project developed and to invite her, the faculty, and the students to future events that the Wave planned to hold. He asked for an e-mail address where he could contact them. Artur had also explained to the principal that the Wave would be building an interactive website where people living near Baikal and Tahoe could communicate with one another to share best practices and to mutually solve local problems. The principal replied that she did not have an e-mail address and that she did not see the point of an interactive Internet site.

"How can I help them develop if they themselves don't want to develop?" Artur asked rhetorically, in a huff. "How can you not want to use the Internet?"

Tanya countered that she herself did not own a television, but Artur dismissed the comparison. "A television is just a television, but the Internet . . . !"

In the evening, we headed toward the village clubhouse for the adult focus group. Artur and Tanya had been on the phone for the past several days recruiting participants. Tanya wanted a minimum of ten people to attend. Tanya said, while we walked down the dirt road, that she hoped the turnout would be better than it had been at the neighboring village, Maloye Goloustnoye, the previous week, when only five people attended. She went on to criticize the villagers as *sovki*—those Russians who are trapped in a Soviet mentality, who expect the government to do everything and show no personal initiative.

TANYA: There was a villager in Maloye Goloustnoye who complained that the fences had not been painted in thirty years . . . So no one has painted the fences—then paint them! They expect everything to be done for them. And that is how it was in the Soviet Union. There was the *leskhoz* [state logging company], and it provided people with homes, it built the school, it gave people vacations. So people came to depend upon the company. It was the whole life of the town, and then it just vanished. The *leskhoz* probably built that fence thirty years ago, painted it, and since the *leskhoz* shut down, no one has touched it.[4]

We arrived at the clubhouse, which was a long wooden building, painted in fading blue and green. Although it was almost 5:00 p.m., when the focus group should begin, the building was still locked. Only one of the invited participants showed up to wait with us by the front door. A town magistrate arrived on a motorcycle with her two-year-old daughter tucked into a sidecar. She opened the clubhouse for us and we all went inside.

The interior of the clubhouse was old and dilapidated. We were shown into a very small auditorium that doubled as the local discothèque. It had red plush theater seats lining the outside walls: some missing arms, others missing seats, some with torn upholstery, and all of it dusty and degraded. There was a proscenium and a stage that looked to be about two meters wide and deep.

We waited past the 5 p.m. starting time, hoping more people would come. Slowly, a few more trickled in. One woman, a recent retiree with short white hair, kept looking around and wiggling in her seat as though at any minute she might get up and leave. It seemed all she could do to sit there and wait. "How can we just be sitting here when there are potatoes to be dug?," she asked aloud to no one in particular.

Tanya was clearly dissatisfied with the low turnout, but tea and cookies could only occupy the guests for so long, and Artur decided to begin. He made a brief presentation about the project that the Wave was undertaking. I had heard his opening remarks shift over the course of the trip. While he had usually begun in a respectful and upbeat tone to express the Wave's desire to help and support villagers, the resistance he had encountered hardened his approach. Now he stood before them and explained his point of view in no uncertain terms:

> The economy is changing . . . either the village will develop or it will dry up. Usually development happens when some fat cat [bogataya dyadya] comes in and pays a ton of money to build something. But we want to know what we can do here with what we have, so that you don't have to wait for the fat cat, who may or may not put money back into the village.

After Artur's introduction, Tanya began her focus group protocol. As the focus group continued, a few more stragglers came in. All told, seven people showed up to the focus group. The last member came thirty minutes into the focus group discussion (and fifty minutes after the originally planned starting time.) She perched on the edge of the chair closest to the door. When Tanya tried to include her in the conversation, she motioned with her hand to suggest that she was only there to observe, not to participate.

Like the youth focus group, there was much complaining about the state of the village, the prospects for development, and a general pessimism about the possibility for improving the situation. The people in attendance had been specifically invited to the focus group because they were considered to be the most "active" members of the community, but even these frequently reverted to a refrain of fatalism.

"The administration doesn't listen to us," said a woman in attendance. "We can't do anything. We are powerless."

The following month, the Wave held another seminar, this time in the neighboring village of Maloye Goloustnoye. The Wave members planned to bring representatives from the Small Business Administration in Irkutsk, the regional capital, to talk about various programs and opportunities that the villagers could access to help develop small businesses. Villagers from Bolshoye Goloustnoye were also invited to attend the seminar, but all declined.

"They said it was too cold to wait for the bus," Katya said.

"They have a point," I replied, as the Siberian winter was now in full swing. She shook her head dismissively. "So we offered to have a van pick them up in front of their houses," she went on. "We would drive around to people's doorsteps and collect them from their homes. They wouldn't have to wait outside at all. But they still said no. Now they said it was because the cow had to be milked in the morning."

The activists from Irkutsk and the villagers from Bolshoye Goloustnoye approached the SEZ with decidedly different dispositions, developed in their own unique circumstances. Village "habitus" (Bourdieu 1977) was characterized by fatalism and apathy. Efforts at creating change were viewed with suspicion and doubt. Even the schoolchildren believed that only the government could bring change to the village, as in the case of the nonexistent preschool. Not only were the open meetings poorly attended, but those who did attend put on a display of disinterest: sitting near the door and complaining that they had much better things to do than discuss economic development. They performed cultivated disinterest to communicate to others their savvy and situated knowledge.

In a small rural village under a government that is alternately authoritarian and inept, skepticism about the efficacy of individual initiative cannot be easily faulted. In fact, it is hardly even surprising. Scholars in the mid-twentieth century even presumed that dictatorships arose from a pre-existing authoritarian political culture (Broderson 1957; Kornhauser 1959; in Russia, see Keenan 1986). Others have acknowledged that the relationship between governmental structure and political culture is not causal; rather, the two are adaptive and mutually reinforcing (Lipset 1960; Almond 1983).[5] For the villagers of Bolshoye Goloustnoye, general neglect of their needs and desires, both from the state and the new market economy, has not offered them much to lend credence to the ideology of individual initiative.

The activists from Irkutsk, on the other hand, brought with them different presumptions: rather than seeing change as suspect, they consider it inevitable. For the Wave, one needed to keep one's ear to the ground, to keep an eye out, to protect nature from those who would exploit it for profit. The bustle of modernity requires constant vigilance, and networks of like-minded supporters are an important weapon for countering the agendas of the powerful. What was perhaps most incomprehensible to the urban sensibilities of

Irkutsk's activists was the disinterest their rural counterparts expressed at joining the global march of modernity. There is probably no tool as emblematic of the cosmopolitan ethos as the Internet, and yet the school principle practically rebuffed it. When discussing the tourist economy in focus groups, most residents viewed tourists with disdain. They were glad of the extra income but expressed displeasure with the myriad ways that tourists disrupted their village life. They would not turn down the opportunity for running water but hoped the price for such luxury would not come at the expense of their daily patterns and practices. Faced with the intransigence of nonparticipation in this rural community, Artur essentially declared to them at the start of the focus group that development was inevitable: modernity was coming to Bolshoye Goloustnoye, whether the villagers wanted it or not. The only question was whether they would be proprietors of their own guesthouses or scrubbing the floors of a large hotel owned by a Moscow oligarch.

The dispositional divide, defined as rural versus urban, conservative versus liberal, or traditional versus modern, plagued the relationship between the Wave and the villagers they sought to recruit and engage. Russian environmental activists were unable to rally Russian villagers. It took a new kind of engagement, in the form of a transnational webinar, to light the spark.

The Sister-City Webinar Project

In January, the Wave and TBI embarked on the second half of their project: a series of webinar sessions to take place between communities in Russia and the United States. The Wave selected two locations for the webinar in Russia: Baikalsk, because of the long campaign against its paper mill, and Bolshoye and Maloye Goloustnoye because of the potential SEZ to be placed nearby. South Lake Tahoe in California would participate because it had a long history in the region as the home of TBI and because it is a designated "sister city" to the city of Baikalsk. Because there was only one computer in Bolshoye Goloustnoye, and the dial-up connection was too slow for video, any interested participants from the Goloustnoye villages had to be picked up in a minivan and driven to the Wave office in Irkutsk for the webinar event. I also attended the first webinar at the Wave office as a participant observer.

FEBRUARY 5, 2013

It was still dark when I arrived at the Wave office at 8:45 a.m. for the very first of the webinar series. Marina and I greeted each other as I shed my layers and hung them on the coat stand.

"We have only two people from Bolshoye Goloustnoye and no one from Maloye Goloustnoye," Marina told me, with a grin. I was perplexed as to how I should react, as I often was when Marina would deliver bad news with a cheerful smile. She then directed me to the classroom in the back of the office. I saw a number of chairs set up in a semi-circle facing a computer monitor, with a microphone and speakers attached. There was an embarrassment of staff in the Wave office: three times the number of employees as there were program participants. In addition to the regular Wave staff, two scholars from the Center for Independent Social Research (CISR) were on hand to assist, and a journalist had been recruited to document the event. Marina encouraged everyone to help themselves to the coffee, tea, muffins, and cookies brought for the occasion.

The two guests from Bolshoye Goloustnoye sat in the chairs, facing the computer uncertainly. One was a woman—Zina—in her late sixties probably, with short, spiked white hair. Beside her sat a man named Pyotr who looked to be in his fifties. His long black hair was streaked with grey and pulled back in a ponytail. He had dark rimmed glasses and a goatee.

Olga, a sociologist with CISR, was serving as moderator and translator. She thanked the two from Goloustnoye for coming and explained to them what the webinar would be about. She told them of the similarities between Baikal and Tahoe. "Thousands of people live in Tahoe, and they get millions of tourists. Two million people live on Baikal, and we get thousands of tourists. But tourism is something that we have in common. So we can talk about tourism here and tourism there."

"It will be interesting to listen to it," Zina told Olga, although she looked very far from interested, her face buried in a gardening tabloid she had brought with her.

"Before we begin, let's talk about the questions that will be raised during the webinar," Olga continued. "What are the key assets or tourist attractions in your region that you can discuss during the webinar?"

The two guests sat mute for a moment. Then Zina shrugged and said, "The Dry Lake."

"What is the Dry Lake?" Olga asked.

"It's a lake in the forest that appears every four years," she answered, and then was silent.

"So, every four years there is a lake. What happens the other three years?" Olga prompted.

"It's just a field."

"So it's pretty?"

"Well, yes, it's pretty."

"Ok. There is the Dry Lake. What else? What other things would attract tourists to Bolshoye Goloustnoye?" Olga asked.

"Lake Baikal," Pyotr said. "In the winter, you can ice skate on the lake." Given his prosaic answer, one might have countered that Baikal is not unique to Bolshoye Goloustnoye. Nor did the village have any particular corner on the market for ice skating. Olga's expression and tone suggested disappointment and mild frustration with their disinterest, but she proceeded with the project program and prepared to begin her role as moderator and translator for the Irkutsk site of the webinar.

The Wave had selected Google Plus as the hosting software for the webinar. All three locations could be seen at once as small inset panels at the bottom of the screen, and users could take turns enlarging them to the full screen size. Olga opened the screen from Tahoe, and the two guests from Goloustnoye inhaled sharply.

"So many people!" Zina murmured, and indeed, with about eight attendees, there were noticeably more Americans seated around the table in Tahoe's conference room than in either Irkutsk or Baikalsk, which had two and four people, respectively. There was also a small squeal of delight when one of the Russians noticed a large map of Lake Baikal on the wall of the conference room in Tahoe.

The conversation got started. We began with introductions, going around to each individual at each site. Dialogue was understandably slow, since everything needed to be translated into either English or Russian by the translator that each site had on hand for that purpose. But despite the continuous need for translation, the conversation soon became more involved.

The Tahoe residents came in with an agenda. They had a model of development that they clearly wished to convey. The head of a local nonprofit, a woman with long silver hair, described Tahoe's program for sustainable tourism.

"Our region is embracing geo-tourism as a way of life," she explained, without defining what geo-tourism is.[6] "First, you have to identify those assets that are unique to your area." Here she produced a map of Tahoe with several points indicated upon it. "Then we develop these places on the ground, so we can show tourists these particular places." Here, she showed information about a number of "adventure tracks" that had been developed for tourists to follow. "We use a trinomic model in developing these tracks that includes social organizations, private businesses, and the government. Nonprofit organizations host the adventure and provide the guides, the state provides access to these special places, and then the local businesses benefit by providing the food, equipment, and transportation. And everything we use in an 'adventure track' showcases the local—it's all local. Local food, local transportation, local kayaks, local businesses, local biologists. Everything is here. Nothing is imported. There are no cars [involved]."

While she listened to the accented Russian proceeding from the translator in Tahoe, Zina nodded her head vigorously. "Visitors have incredible experiences," the woman in Tahoe went on. "It's fun, it's educational, and it's low-carbon . . . It's sustainable."

Next, the conversation turned toward tourism potential in Baikalsk. There was a doctor in attendance, employed by the local ski resort, who harbored dreams of developing medical tourism in Baikalsk through alternative spa therapies. The Baikal watershed is seismic, with hot springs throughout the region. This doctor planned to develop thermal springs therapy and blue clay spa treatments with local muds in Baikalsk. He presented this as a means to sustain tourism outside of the regular ski season, but noted that it would be a major economic undertaking and would require a great deal of collaboration and capital to succeed.

After hearing about the well-developed program of adventure tracks in Tahoe, and a thermal spa in Baikalsk, it was time for the two villagers from Bolshoye Goloustnoye to speak of their own tourist attractions. They spoke of the Dry Lake and the skating trails, as they had with Olga before the webinar; however, this time they did so with more enthusiasm. Zina described the Lake as a magical, mysterious place that everyone should come and see. She talked about its healing properties and the legends surrounding it. She seemed to think the site more worthy now than before the webinar. Pyotr talked positively about the idea of creating an ice skating trail, and continued to speak about the potential of ice skating tracks even after the conversation had turned to the difficulties of tourism and the problems it can bring. Their pessimism had morphed into interest, positivity, and participation.

The conversation proceeded nicely until this point, and it seemed that the three sites were glad to listen to one another's issues and ideas. But discussion then took a sudden turn for the worse. One of the predesignated questions asked each community to discuss various environmentally friendly technologies that they used in tourism, and the drawback to cross-national collaboration became immediately apparent.

"Embassy Suites is a four-hundred-room hotel in South Lake Tahoe," explained one of the community leaders there. "They managed to cut costs by $500,000 per year based upon environmental upgrades." She explained that the hotel now composts all its food waste, and how it instituted a recycle-reuse program that helped save money.

"But one of their biggest expenses was laundry," the woman went on. "So they bought new, efficient, ion cleaners, which are modern washing machines that use little energy, little water, and no soap. They use no soap and no dryers!"

The participants in Russia looked back at the screen blankly. The "best practice" offered by Tahoe in this instance was so divorced from their own experience that the conversation seemed to shrivel and die. The villages of Goloustnoye have no plumbing—the schools and hospital have outhouses.

Washing machines themselves were fanciful, let alone ion cleaners. Even in Baikalsk, which is an industrial city, one can see clothes hanging out to dry in subzero temperatures.

Someone in Baikalsk muttered something about a solar panel somewhere in the city.

"Our technology is simpler," Zina answered when it was her turn to talk about "alternative technologies" in Bolshoye Goloustnoye. "Our leftover food is given to the cattle. And we use only fallen trees for heating the house, rather than cutting trees down."

No one in the webinar was directly helping anyone else at this moment in any way. Bolshoye Goloustnoye had no use for Embassy Suites and its ion cleaners. People in Tahoe only heat their homes with fire for the aesthetic, and they certainly would not be giving food waste to their nonexistent livestock. Baikalsk hovered somewhere in the middle. As a factory town, they had central heating, so felling trees was not a chief concern of theirs. But neither would their solitary solar panel do much to impress the Americans.

To bring the conversation back onto mutual footing, one of the Russians in Baikalsk asked the Tahoe residents to tell him about snowshoeing, which is a sport seldom seen in Russia. The Americans obliged, talking about their adventure tracks and how local businesses will rent out snowshoes.

"What about ice skating in Tahoe?" asked Pyotr, seeking to include Bolshoye Goloustnoye in the conversation.[7]

"Tahoe never freezes," the Americans answered in one voice, resulting in raised eyebrows from the Russians who live by Baikal, where winter ice is more than a meter thick.

"We have ice skating in Truckee," said one woman referring to a city not far from Tahoe. "But it's indoor."

"Come to Bolshoye Goloustnoye!" the two residents invited their foreign correspondents with enthusiasm. "We have the biggest ice skating rink in the world!"

And with this friendly invitation extended, the first international webinar between Baikal and Tahoe drew to a close.

Bolshoye Goloustnoye had a reputation among the Irkutsk activists involved in the project for its passivity and intransigence. Invitations to participate in events were inevitably met with reluctance and questions like: "Why are we having a meeting when the potatoes need to be dug?" or "How can we travel to Irkutsk when the cows must be milked in the morning?" There were a few active members of the community, but the general public in this village of six hundred souls tended to enact an almost ritualized avoidance of involvement. It was that performance of disinterest that brought Zina to press her nose to her magazine and to tell Olga that the conversation would be interesting "to *listen* to," as though to emphasize that she was there to hear and not to participate. She would attend but keep herself at a distance. Such insouciance is also the likely reason that there were so very few people from Goloustnoye in attendance for the first webinar.

But during the webinar, there was a marked change in the participants from Bolshoye Goloustnoye. Their cultivated disinterest failed them when faced with the active interest of Baikalsk and South Lake Tahoe. In the village, one garners a degree of respectability from suspicion and stubbornness, but during the webinar, respect was conferred on those who were progressive: to those who were attempting to improve their communities—and especially to those who had succeeded in some small way to make change. Suddenly, participants from Bolshoye Goloustnoye felt the need to tout their potential skating trail, to talk about their Dry Lake with more gusto. It was the only thing that saved them from the mortification of having nothing to say when, in a new social milieu, the standard for respectability had shifted dramatically *toward* involvement.

However, the transformation that occurred among the participants in Bolshoye Golousnoe was not merely one of lip service, nor was it solely geared toward saving face, as I thought at first glance. Instead, it seemed that simply being exposed to another context, one where progressive thought was validated, altered the apparent *legitimacy* of activism and community involvement for the villagers. Both participants listened attentively, and Zina would nod her head on occasion, signaling that she heard what the other participants were saying, and that she liked what she heard. By the end of the conversation, both participants from Bolshoye Goloustnoye were fully engaged, offering an enthusiastic farewell.

After the webinar, Zina behaved differently. She had long since tucked her newspaper away, and now she wanted to talk, to work through the ideas and the experience. The two guests from Bolshoye Goloustnoye showed a reluctance to leave the Wave's office after the webinar. They wanted the conversation to keep going and began telling the staff what they thought of all they had just heard. Indeed, the desire to debrief was so evident that the Wave decided in future webinars to formalize it and hold a post-webinar discussion as part of the process.

Importantly, the effect of this cognitive transformation extended beyond the webinar itself. The sense of interest and efficacy did not stop at the door of the Wave's office. Reports of the webinar spread, and at each subsequent meeting in the series, more and more people were in attendance on the Russian side of the screen. There were six webinars[8] altogether that took place in Irkutsk, Baikalsk, and South Lake Tahoe. Each saw increased attendance on the Russian side of the screen through the penultimate webinar.[9] Participation rose from two to twelve, and, while a dozen locals may not portend a social revolution, it does represent a sixfold increase over the course of a month, and it was in total opposition to the Wave's previous experience recruiting in Bolshoye Goloustnoye. The growing interest and widening participation suggests that the webinars were moving the thought processes of the villagers toward more willing involvement.

In addition to raising interest in participation, the project also influenced people's sense of efficacy in the face of social problems. The webinars became spaces where stumbling blocks that had seemed insurmountable became collectively conquerable. Once in a mindset of efficacy and agency, problems got solved. During the webinar that discussed poverty, a woman with a guesthouse in Bolshoye Goloustnoye, Galina, talked about putting in a new pit toilet and then being fined 130,000 rubles (US $4,334) because she did not have proper documentation for it. She described her difficulties with the state bureaucracy that regulates business and concluded, saying, "I'll have to work on the black market, because I just can't deal with what it takes to work officially."

This type of comment is common in Russia. Anthropologist Nancy Reis (1997) refers to this discursive category as a "lament." In her ethnography of discourse in the late Soviet period, she documented the cultural tendency to describe at length the problems and difficulties people face, particularly at the hands of an incompetent and unresponsive state. An important quality of the "lament" was its divorce from any discussion of problem *solving*. This litany of suffering was not geared toward finding an avenue to relieve the suffering—it was simply a discursive ritual: to proclaim a lamentation. The lament has continued to persist in Russian culture well past the *perestroika* period wherein Reis documented it, particularly in the smaller towns and villages. Normally, Galina's complaint would have fallen squarely in this tradition.

In the context of the webinar, though, the lament was met with a nontraditional response. The villagers began to try to help her overcome the obstacles she described. They asked her questions about how her business was classified and made recommendations for what she could do. Among those present was a woman from the Irkutsk Small Business Administration, and she passed her card to Galina, saying she would be glad to help her work out her documentation problems and point her in the direction of grants and resources geared toward small entrepreneurs to make the toilet and other improvements more affordable. The two met up and continued talking after the webinar.

In another occurrence, during a post-webinar debriefing, participants from Bolshoye and Maloye Goloustnoye began to discuss about the growing litter problem that has accompanied the rise of Baikal tourism.

"I'm ashamed of the mentality in our country," another woman piped in. "We have volunteers come to pick up litter. And not just volunteers: international tourists will pick up litter and bring it to me. But our Russian people just drink and toss the bottle."

Again, despite the tone of the comment, which suggested a lament (and to which the appropriate response is a sad shake of the head, a click of the tongue or another lament), those at the webinar began to think of ways that the problem might be fixed.

"My souvenir shop could start a program: bring a bag of trash and get a free souvenir!"

The webinars provided a space apart from the everyday where possibilities could be envisioned and expanded. In the third webinar on poverty, a woman who owns her own restaurant in Baikalsk attended.

> "We started with four people, and we now employ between seventy and one hundred people, depending on the season. It is hard to do business because of high taxes and expenses, but we pay good salaries. We also do charity work. We built a church with our contributions. We also regularly provide free meals to children in a local school. We are Russian, we are a strong people, and we just have to work [and] to believe in our work."

At the end of this speech, the participants in Irkutsk from Bolshoye and Maloye Goloustnoye burst into applause. They had been listening with rapt attention throughout, exchanging nods or approving glances with every point the speaker raised. They were impressed, not only with the success of the restaurant, but especially with the apparent civic virtue displayed by the local entrepreneur—providing wide employment, with a "good salary," feeding local school children for free, and building a community church.

During the group discussion following the webinar on poverty, the very first comments were about the restaurant.

"Let's organize an exchange so we can meet in person," someone from Bolshoye Goloustnoye said. "At least between us and Baikalsk. We can send delegates there, and they can come here. Especially the woman with the restaurant." The others nodded energetically and added that they also would like the opportunity to see the restaurant and learn more about it. For a population that in previous circumstances would find a multitude of excuses to avoid attending seminars or meetings, it was a radical departure for the same group to suggest such an exchange of their own initiative. For a group that would often complain of the difficulties in traveling to Irkutsk, or even the next village over, the excitement about a field trip to distant Baikalsk was out of the ordinary.

Why did the webinars succeed in fostering some movement toward participation and problem solving when the Wave's own attempts had not? The crucial difference seems to be the transnational aspect of the project. But is it transnational engagement specifically that causes this change? What was the apparent impact on the Tahoe side of the screen?

Consuming the Other

Not all parties came to the webinar with the same expectations, and these may have conditioned the impact that the webinar had in each community. Social position can temper the outlook of unequal parties in a dialogic exchange. The difference in expectation between participants in Russia and the United States was most evident at the fourth webinar, which covered the topic of food.

When the Tahoe participants discuss food, they are chiefly concerned with issues of environmental protection, and it derives from the American intelligentsia's growing distaste for agribusiness, with its use of chemical pesticides and herbicides and poor treatment of livestock (e.g., Pollan 2006). Moreover, many US environmentalists are worried about the carbon footprint of their diet. "Food miles" is a term used to encompass the distance traveled by the beef or broccoli or rice that ends up on a person's plate (e.g., Weber and Matthews 2008). To counter the negative effects of agribusiness and food mileage, a movement has grown up in cities in the Global North to bring food production closer to home through farmer's markets and community-supported agricultural (CSA) cooperatives. Such were the food concerns that motivated the webinar participants at Lake Tahoe.

Like the United States, Russia also developed a form of industrialized agriculture, but there remains a major difference between the two countries when it comes to the production of food. The practice of domestic agriculture was not lost in the Soviet Union as it was in the United States. Collectivization and food production *po planu* (according to plan as opposed to the market) resulted in problems that ranged in severity from famine to waste, from scarcity to spoilage. To address food problems and to serve as an incentive to Soviet citizen employees, state companies began to distribute a parcel of land to their workers: a *dacha*.[10] Usually an hour or two from the city by commuter rail, these rural plots would host a primitive cabin and garden that would grow flowers, vegetables, berries, and fruits. Soviet Russians would spend their summer evenings and weekends working the land, and the yield would produce enough both to feed themselves for the summer and to preserve food for the winter.

About 40 percent of urban households possess a dacha (Clarke 2002). Thus, even urban residents maintain the art of subsistence agriculture. Dachas ensured fresh food supply during the vagaries of planned production or periods of economic crisis in the market economy. The extent to which Russians make use of subsistence agriculture generally follows the rise and fall of economic well-being in the country as a whole (Southworth 2006). Residents in Baikalsk and Bolshoye Goloustnoye are no different in this regard. When the paper mill suspended its operations in 2008, many people relied upon selling dacha-grown strawberries[11] on the side of the road for income. While villagers of Bolshoye Goloustnoye had always maintained their own agricultural plots, these were not their main form of subsistence during the Soviet years, when most residents worked for the *leskhoz*. But after Bolshoye Goloutnoe was incorporated into Pribaikalskii National Park, the logging industry shut down, and what had been the villagers' supplementary agriculture became their chief occupation.

When the Russians and the Americans came together to discuss food, their unique national histories limited the kind of assistance that could be offered. The Tahoe interest in food was philanthropic and philosophical, but it was not existential. Their approach to food was that of a conscientious consumer. For the Russians, self-sufficient food production was considered both necessary and good. Given the short growing season, Russians are also accustomed to buying food that has traveled many miles. Any fresh produce purchased in winter has been shipped in from China or Central Asia—and with winters that can reach forty degrees below zero, there is no easy way to shorten those food miles. Instead, Russians emphasize that, while it is necessary to ship food in winter, one's own produce is always preferable.

Given their preoccupation with food miles, the participants in Tahoe began the webinar by stating their desire to grow food in Tahoe and acknowledging the impossibility of such an endeavor. "Almost no one up here grows any food," one Tahoe resident explained. "Quite a few people have tried to grow veggie gardens, but they don't succeed." The residents of Tahoe were convinced that the heavy environmental regulation protecting the lake from erosion and runoff would render domestic agriculture impossible. But they went on to say that Tahoe makes up for this lack with farmer's markets and CSAs, where food only travels fifty to two hundred miles.

Russian participants were pleased to learn about CSAs and ways to organize the sale of rural produce to nearby urban dwellers. However, they also wanted to share their own knowledge of domestic agriculture with the Americans in Tahoe. Russians in both locations talked with enthusiasm about their fresh eggs, chickens, berries, and vegetables. Residents of Baikalsk even brought examples of fish, nuts, and homemade jams to show proudly to the Americans in Tahoe. They explained how to start seeds indoors in late winter to circumvent the short growing season and how to preserve freshly grown produce to eat in the winter.

One man from Baikalsk discussed his permaculture garden, which he framed as useful to the residents of Tahoe and their environmental difficulties:

> We have the same problems as Tahoe, with little top soil and a cold climate. But I read an American magazine a few years ago about gardening, and there was all this information on permaculture. Now I have an organic garden, making use of an organic landscape design. Fruit trees are not common in Siberia, but I have managed to grow apples and pears, apricots and cherries. They produce fruit for us, and more left over, so we can feed our friends. I also use the garden to educate people about organic gardening and permaculture. I invite people from the local campgrounds and hotels. Every year, the students at Tahoe-Baikal Institute come as guests to my house and tour my garden. We also grow saplings and flowers that are used for landscaping in the city. We do all of this organically, and we only have 1/3 of a hectare [less than an acre].

This man combined the Russian tradition of domestic agriculture with ideas on permaculture that he imported from an American magazine. His experimentation had proven successful, and he was eager to share his knowledge and experience with the representatives in Tahoe. However, they did not seem interested in learning from Russia's experience. Despite their rare access to expert domestic gardeners and subsistence farmers in a similar ecosystem, they continued to write off the very possibility of homegrown produce. The usefulness of this best practice was as lost on their sensibilities as ion cleaners had been for the villagers in Russia.

Instead, the food webinar served to reinforce the axiomatic expression of multicultural appreciation: an exchange of recipes. In multicultural America, the exotic is domesticated and incorporated via gastronomy (e.g., Appadurai 1988). Urban dwellers can display their worldliness and tolerance through adventurous cuisine. They can imagine themselves partaking of a culturally authentic experience when ordering Vindaloo or Pad Thai, from the safety of a restaurant that is expressly created for the purpose of Western consumption (Long 2004). So too did the Tahoe residents seek to display their appreciation of Russian culture by reducing it to food. The one question that Tahoe residents asked of Russians with the intention to learn a "best practice" from them was: How do you cook cabbage? "We would like to know how to cook cabbage," a woman in Tahoe asked. "I think cabbage would grow well here, but Americans don't eat cabbage, mainly because we don't know how to prepare it." Were she ever in Russia as a guest or tourist, she said, she would like to eat Russian cabbage.

The grant project that supported the webinars was based upon the principles of exchange and cooperation. But as the project progressed, it became increasingly evident that the concept of mutual benefit was not itself mutual. American participants were glad to talk about their own work, and they were always glad to hear what their colleagues in Russia were doing. But there seemed to be an implicit expectation that one party was there to provide the best practice, and the other was there to receive it. Whether it was the trinomic model of geo-tourism or the establishment of CSAs, Americans came prepared with programs to impart. There was less evidence that they had any expectation of *receiving* help from Russia. While that may be an accurate assessment, given the different levels of development between the two regions, the very assumption may have forestalled learning; it may have prevented a similar expansion of mental horizons in Tahoe, such as happened in Bolshoye Goloustnoye. While transnational activist collaboration was able to bring to rural villages the experience of cosmopolitanism, the routinization of cosmopolitanism in the developed world prevented them from experiencing the opportunity for wonder that comes with encountering difference.

What Is in It for Them?

While webinar participation in Baikalsk and Irkutsk continued to grow, it was waning in Tahoe. It became clear that the participants on the Tahoe end of the conversation were almost exclusively composed of individuals who spearhead local nonprofit organizations and who were invited to participate for one particular subject of discussion. Certainly the same was true for individuals in the Russian sites as well—the man with the permaculture garden did not come to the food webinar by chance—but there were also those people who kept coming back to the webinars regardless of the subject matter. These were people eager to learn what was going on in other locations and excited or inspired by the ideas that were shared. In Tahoe, it seemed that the only repeat attendees were the organizers from TBI, and this gave the impression that while the Americans were glad to talk about their own programs, they were otherwise uninterested in their sister lake across the ocean. The lack of comparable "repeat attendees" on the American side was noted by Russian participants with concern. Were the Americans engaging in exchange, or was it merely a paternalistic charity they offered, sharing their assumed greater wisdom with their imagined inferiors?

The penultimate webinar discussed the problem of retaining young people in the towns, of developing activities and employment opportunities for youth so they would not move away to larger cities. With eleven locals present, this session was the most populated of all the sessions in Baikalsk, and was attended by young people themselves who had learned about the webinar from their teachers or parents. One attendee was a pretty blonde girl in her late teens. In the post-webinar discussion, she said that she wished students in Tahoe had attended the call on the American side so that they could have talked. Everyone else nodded and agreed.

"We had young people here, why didn't they have young people there?" asked one of the adults with a critical tone, who had been a repeat participant in the webinars. "They just brought their NGOs and teachers. If you want to talk about problems of young people, then you have to talk *to* young people."

Again, in the spirit of the webinar and its active, problem-solving, and initiative-seizing ethos, participants began brainstorming ideas to connect young people in Tahoe and Baikalsk. Someone said that a youth exchange should be part of the sister-city relationship. Others thought it would be cheaper to have a display in each of their schools where they could post photos and stories about one other. One popular idea was to have groups of students in each village plant a tree on the same day in honor of each other, then send images and information about the trees back and forth as they grew: that way, students would be caring for nature locally while imagining their brethren globally.

While the post-webinar discussion carried forth the optimism and initiative that the webinar space tended to foster, doubts lingered over whether the wealthy, progressive, democratic Americans were actually interested in their Siberian counterparts, other than as a fleeting curiosity.

FEBRUARY 25–26, 2013

Our hotel in Baikalsk looked like a warehouse from the outside. The inside did little to mitigate the resemblance. The lobby was white and open, only sparsely furnished with a sofa and concierge desk. The lights were kept off in any part of the hotel that was not in use, and the corridor that sank back from the lobby was black and forbidding. After checking in, it was down this corridor we went, stopping at the one door that radiated light from the inside. The room contained four twin beds with mismatched sheets in flamboyant designs.

"Ooo, I want the zebra bed!" Katya announced with enthusiasm, claiming the animal print sheets as her own. The rest of us lumbered in and dropped our belongings. We pulled two chairs together and improvised a table to quickly eat dinner. We had homemade *blini* [Russian pancakes] with ground beef and rice stuffing, a spicy carrot salad out of the jar, sesame seed bars, and bread with jam. For dessert, Katya bought cookies that were molded into the form of cats.

"I thought you were a vegetarian," Artur teased her, as she gleefully bit the head off one of them.

It was in this hotel that the Wave was holding a closed meeting in preparation for the last videoconference in the sister-city project. The last webinar was supposed to discuss the "sister-city" relationship between Tahoe and Baikalsk, and how that relationship could be reinvigorated. The Wave invited select individuals who had been active in the formative years of the "sister-city"[12] partnership between South Lake Tahoe and Baikalsk. When visits to Russia were cheap and grant money flowed plentifully to help integrate societies long separated by an Iron Curtain, there were repeated delegations sent in exchange between Tahoe and Baikalsk. The alumni of these various exchanges gathered on the second floor of our small, warehouse-style hotel in Baikalsk to discuss that period of time.

The second-floor lobby where the meeting took place was cold and dimly lit, but was very large with a high ceiling. We set up chairs to form a circle, incorporating the couches and armchairs already in the pink pastel room. Guests started trickling in. There were twelve individuals from Baikalsk, joining Marina Rikhvanova, Katya, and Artur from the Wave. At first there was simply pleasant discussion and reminiscences about past exchanges. Then Marina Rikhvanova brought the group to order. She organized the conversation around four questions: what they did in the past, how they had benefited, why the relationship ended, and what they might like to see happen in the future. For each question, she posted a sheet of flip-chart paper and stood ready with a magic marker to list the group's responses.

The members described four exchanges that took place. One was geared toward school children, one was for city administrators, and two were professional exchanges, including for local artists, two of whom were among the twelve people present.

"It was the 1990s, so people had never seen anything like it," said one of the men. "No one went abroad back then."

"I remember how shocked those Americans were when they saw Siberia for the first time," said one woman who had hosted exchange guests. "They learned three new words: *salo, vodka* and *moroz*! [lard, vodka and frost!]"

"I also remember the shock *we* had when returning to Russia after our trip to the United States," a soft-spoken man added.

"Ugh! Yes," another woman chimed in. "The airport! As soon as we arrived in the airport in Russia, there was this smell, and everyone was dressed in black, and no one smiled."

"I remember when the school kids came back, they came to me because I worked in the city administration, and they asked me, 'Why is it not like that for us here?'" The small woman with short hair and glasses seemed overcome with emotion remembering those young people who wanted answers to questions about their country's unequal development. Marina turned to the posted paper and wrote: "Kids see the world," on it under the list of benefits resulting from sister-city exchanges.

"What were the other benefits of the exchange?" she asked the collective.

"Learning a foreign language," one participant said.

"It improves the image of the city," said another. "Baikalsk could be known internationally."

"And not just as a place with the smelly paper mill!" the black-haired woman added.

After a pause, one of the artists spoke up. "It created a new level of discussion," he said. "It fostered a more intellectual level of discourse."

The other artist, a thin man with a bristly brown moustache, then spoke up. "And there were even financial benefits. Several of my colleagues and I showed our work at a gallery in America and we even made some money on the event."

Next, Marina Rikhvanova asked what happened to the sister-city partnership. "Administrations changed," several people said at once. When asked for clarification, they said that when Vladimir Putin and George W. Bush each came to power in 2000, their governments set different priorities than those of their predecessors. Also, the terrorist attack on September 11, 2001, in the United States altered funding priorities away from former Soviet countries and toward the Middle East. Visas became more difficult to procure. It was not an abrupt rupture, they explained. Rather, the infrastructure that allowed for the possibility of exchange had gradually eroded.

Then, Marina asked what they would like to see out of a renewed sister-city relationship, and people made many suggestions of various activities that they would enjoy. But when Marina asked for more concrete steps that individuals might take to bring such activities to the next level, the conversation became stilted. The hesitancy did not result from laziness, lack of faith in efficacy, or an unwillingness to take responsibility for implementing the projects. Instead, it was tied to a concern among those present that the Americans in Tahoe were actually uninterested in a long-term relationship with the people of Baikalsk.

"What I want to know is: do they, in Tahoe, really want brotherhood, or do they just want to look at each other on the computer screen?" asked the small woman who used to work for the city administration.

"Is this conversation happening over there, too?" another man wanted to know.

The people of Baikalsk were unwilling to get their hopes up or to put effort into continuing the project if the Americans did not also take it seriously, and the Russians were skeptical that the people in Tahoe had the interest or ambition to do so.

For the residents of Tahoe, the sister-city relationship with Baikalsk in Eastern Siberia is only one of many ways that their community is linked in to the global imaginary. One group of Tahoe residents is firmly committed to Lake Baikal and to their brethren in Eastern Siberia, and that is TBI; but for others in the community, the link is barely noticed. Tahoe does not need recourse to a sister-city relationship to affirm its cosmopolitan identity. For Baikalsk, on the other hand, it *matters* that they are known abroad, and that they have a tenuous lien on South Lake Tahoe in the symbolic relationship of sister city. This connection provides hope for a wider horizon for themselves, their children, and their community. But from the late 1990s until the webinar project, this symbolic capital had not been tested to see what returns it might bring. Connecting to Tahoe via the webinars enabled the local residents to dream big, but the reality of their unequal status—given Tahoe's routinized cosmopolitanism—threatened a rude awakening. They were loath to learn how little their sisterhood was worth.

So What Is the Point?

After the final webinar, members of Baikal Environmental Wave gathered for lunch in the cozy upstairs kitchen in the Irkutsk office. Seated behind the heavy hardwood table, the Wave members talked about a concern that was repeatedly raised by the residents of Baikalsk: Were the Americans in Tahoe really interested in collaboration? It was a question of collective self-esteem for them: people in Baikalsk did not want to put their hearts and minds into a cooperative project only to have the Americans reject or neglect them.

"They are worried that they will work toward creating a partnership, only to find that the other side doesn't want a partnership with them and isn't going to work for it," Artur explained to the group at the table.

The Wave members sat in silence, each lost in thought.

"So what *is* the point of including America?" I asked.

The members exchanged glances, waiting to see who would speak first.

"It seems to me that the most important thing is to unite Baikalsk and Goloustnoye," Artur said. "America is the catalyst to bring them into dialogue. They come because they are interested in talking to America, but they end up talking to each other."

Elena Tvorogova spoke up next.

The point is to see that there are problems everywhere and to see how local people solve their problems [differently]. To paraphrase the great Lev Nikolaevich Tolstoi, we think all happy families are the same, but our unhappy family is totally unique. We say, "Oh, in America everything is perfect, everything is fine, but *we* have all the problems." Once you interact with people in other places, no matter if those places are thousands of kilometers away, you see they have their own problems. But you also see what they are doing to fix those problems. And it gets people thinking of other ways they can be fixed. In a lot of these villages, you see people stuck in problem-thinking. They just see: we have problems, they are awful, these people are to blame, and so forth. If you are sitting, stuck in a problem-mindset, it becomes a vicious cycle and you get nowhere. The key is to move people away from problem-thinking and toward constructive-thinking. So they say, "Ah! We could try this. Let's start a small business that fixes this problem. Or let's start a group working to defend our rights," or whatever. But the important thing is that they are thinking constructively, not just waiting for manna from heaven. When you look at other people's problems and see what they are doing [about them], it can be inspiration. You might think, "I wouldn't do it *that* way, I'd have done *this*." But you are already thinking now in a different way, that you would not have done otherwise.

In answering my question, Elena Tvorogova cogently described what might be considered the most fundamental benefit of transnational activist collaboration. While activists may indeed benefit from foreign financial resources, organizational support, or an opportunity structure that is translocal, these are all, fundamentally, strategic means to an end—geared toward achieving some predetermined collective goal. The webinars suggest, however, that there is a deeper contribution that transnational activism can make, one that is more than a means to an end, but rather is an end in itself. *Transnational connectivity can alter how we think.* It can spur creative thought, changing our approach to social problems and the means for solving them. It helps to overcome the limits to imagination and expands perceptions of the possible.

Conclusion

Baikal Environmental Wave failed to spur engagement in Bolshoye Goloustnoye when it went there on its own, but transnational webinars succeeded in drawing people out and expanding their sense of efficacy. However, we must be careful about how much credit is given to transnationalism itself in creating this effect, lest we fail to recognize the deeper mechanism at work.

The Russian literary critic Victor Shklovsky (2009) coined the term "defamiliarization" to describe how an author can make the familiar seem strange in order to alter the reader's perception of everyday events. Transnational activist collaboration can be seen as a process of defamiliarization. Familiar social problems viewed through the eyes of a foreigner, who is facing similar problems in a different context, can alter the apparent *permanency* of one's own environment. Defamiliarization can then spur inspiration and creative thought, in such a manner that individual efficacy becomes plausible, and eventually possible.

Defamiliarization adds a new voice and a different dimension to our understanding of the role of "the Other" in social science. Most often, a society that encounters an Other uses it a means to define and affirm themselves. "The Other" is set up as an opposing group, often embodying characteristics that the in-group views negatively, and in the process, the in-group builds its own identity (e.g., Said 1978; Bhabha 1984). Encountering the Other is a self-affirming and boundary-defining event. The experience of the webinar turns this literature on its head. Undoubtedly, all the previously discussed processes still stand, but they are incomplete, because the presence of the Other has still another implication that cannot be readily ignored. Essentially, exposure to the Other, renders Otherness in one's own circumstances plausible.

There is a spectrum of distance that appears to determine how a group experiences social difference. When activists from the Wave would reach out to villagers in their own region, the endeavor was reminiscent of the Russian Populists, or *Narodniki*, from two centuries prior. These idealistic intellectuals went out to the countryside to entice the peasantry to help overthrow the monarchy; they similarly foundered upon entrenched conservatism and a rural resistance to democratic progress. Urban elites, whether they are the Populists of yesteryear or the Wave activists today, are different in disposition from their rural counterparts, but it is not perhaps sufficient difference to inspire defamiliarization. The familiarity of common cultural norms overcomes lesser differences in habitus, and the similarity that surrounds the difference breeds skepticism and hostility rather than openness to possibility. Similarly, the webinar discussions suggest that cultural differences must not be too wide, or potential possibility again turns to hopelessness. Ion cleaners may have come from another planet, so much was their irrelevance to Bolshoye Goloustnoye. There is a "goldilocks" space of difference to spur defamiliarization; too close or too far will render the interchange inconsequential.

Cultural difference is not the only obstacle to defamiliarization. Preconceptions may forestall the imaginative expansion that difference can bring.

In this manner, the developed world is at a decided disadvantage. Groups occupying a more privileged position in a globally stratified system may assume they have nothing to learn from exchange and thus close off the opportunity to alter their thoughts a priori. As residents of a globally dominant culture, the Americans in Tahoe find their presumptions reinforced in their daily lives precisely because the terms of global society are set by the West. They came to the webinar prepared to give but not to take. The experience of privilege props up the presumption that one's local practice is already the best practice, and difference is discounted accordingly.

However, it is possible that this predisposition to closed-mindedness in the developed world could be moderated by the duration of the defamiliarized contact. In my interviews with TBI's exchange alumni, Americans were always able to think of some contribution that Russian environmentalism could make to American environmentalism, even more so than Russians would themselves. Foreign GBT participants with whom I spoke also described positive attributes of Russian culture they would gladly adopt and ways their minds were changed by their experiences in Siberia. Extended encounters likely deepen the experience of difference beyond the surface effects that define so many aspects of "banal cosmopolitanism" (Beck 2006). When one encounters a dizzying array of disparate cultural trappings in one's mundane life, as may take place in wealthy, networked Tahoe, exposure to difference becomes so routinized that difference itself becomes normal. One would require a deeper and more subtle appreciation of another culture to actually find one's thinking transformed.

But for those occupying positions in the periphery of the globalized world, transnational activist collaboration could be one of the rare opportunities to experience that which is taken for granted in more cosmopolitan centers. For them, exposure to transnational activist networks could have a profound cognitive effect. In the case of the Wave's webinars, the effect is likely to be only a brief one—a blip on an otherwise static line of doubt. The webinar created a space apart from the local structures and conditions that fostered suspicion and surrender. Returning to the village when the webinar series was ended would likely be sufficient to again restrict one's activist imagination. It would take prolonged involvement with the Other to truly change ingrained habits of thought and culture, to counteract the lived experience of societal limitations (Brown 2016). But the experience of the webinar shows how impressive that change could be if sustained into the future.

And we need not look only to the webinar to draw such a conclusion. The group that the Wave gathered in the Baikalsk hotel to discuss the "sister-city" exchanges fifteen years prior had all been transformed by their own cross-national engagement. These individuals remained some of the more active and progressive residents of Baikalsk over the years, and their participation in the sister-city exchanges likely supported this outcome. The experience of transformative events, even of relatively brief duration, may yield lifelong consequences (McAdam 1988). The webinars themselves did not achieve this, but other transnational activist collaborations might.

Finally, there is one last limitation that the experience of the webinars suggests for considering the mental effect of transnational collaboration: sincerity matters. To sustain the positive outcomes, the engagement must be respectful, open and actively sought by both parties. As the Tahoe-Baikal webinars progressed, it became increasingly evident that one of the two parties was minimally invested in the partnership. When that happened, the other side began to withdraw. Transnational collaboration comes with a risk. The burgeoning faith in oneself and willingness to become involved may result in failure—particularly in those contexts that are prone to produce fatalism and apathy. Local ridicule or state persecution will be difficult to withstand without strong moral support from one's transnational peers. If the investment in the collaboration is not pursued with sincerity, then the result may be disappointment and disillusionment, even worse than before.

But should the sweet spot in the spectrum of difference be found, and should the pitfalls and limitations of collaboration be avoided, the act of reaching across borders and working cooperatively with different cultures could have a uniquely positive effect: increased ability of individuals to imagine social change and their willingness to work toward it. This possibility has vital implications for the field of power. In an era defined by globalization, where people are becoming interconnected as never before, opportunities for defamiliarization are bountiful. In precisely this context, civil society may find its own generalizable power—voluntary action—newly expanded, even as other players in the field of power likewise leverage the global to their own advantage.

Finally, the potential of defamiliarization through transnational collaboration to spur hope and action gives a new metric for valuating local NGOs, such as GBT, TBI, and the Wave. Despite the difficulty that domestic NGOs face in trying to produce change in their home environments, the importance of their continued presence locally cannot be overstated. NGOs should not be judged solely by the success of their particular campaigns or actions, but also by the opportunity they provide to their communities to immerse themselves in transnational discourse, if only briefly. GBT offers this opportunity to its volunteers, TBI to its students, and the Wave to the villagers and town residents outside of Irkutsk through projects like the webinar. These organizations are nodes in a social network existing at the nexus of two planes—the local and the global. Their persistent presence in a particular locality offers that community a continuous portal to the wider world. Individuals may drift into and out of contact with this portal; people may have a life-changing, thought-altering experience as a result of that contact, and then carry the experience forward into their future lives, whether or not they continue to associate with the NGO itself. Nevertheless, the fact that this opportunity was there and available to them matters. When we evaluate civil society organizations, and the service they provide to their communities, their successes may transcend their stated goals. Their deepest contribution may be as unintentional and underappreciated as a sparsely attended webinar. Simply by opening the space for sharing, these organizations help lay the foundation for social change, even in the smallest villages of Eastern Siberia.

5

Putin's Favorite Oligarch

DERIPASKA: *The Russian consumer market is growing and we can sell as much as we can produce.*

SARAGOSA: *Do you ever worry about being too big?*

DERIPASKA: *We're investors.*

SARAGOSA: *At what point do you say that's enough?*

DERIPASKA: *Not at the moment.*

—Interview between Oleg Deripaska and Manuela Saragosa
of the BBC World Service, May 9, 2013[1]

This summer I re-read again Atlas Shrugged *by Ayn Rand.* Atlas
Shrugged *is a legendary series ... In the US the series is regarded as
the second most important after the Bible. This masterpiece made Ayn
Rand a global literary star.* [2]
—Oleg Deripaska,
on his website regarding books he recommends

Certainly transnational activism has profoundly impacted the post-Soviet development of local environmentalism around Lake Baikal. But activists were not the only groups in Russia at this time that were engaging with their global peers and transforming themselves in the process. Economic elites, who wield the generalizable power of money, also entered the field of power in Russia after the Soviet collapse. They too began to engage with the global market and were likewise involved in learning and adapting to a broader horizon. In so doing, their relationship to other actors in the field of power necessarily changed. This chapter explores the interrelations of corporations and local environmentalism—holders of financial and civil power, respectively—through the person of Oleg Deripaska and his private holding company, the En+ Group, who have become inextricably interwoven into the local scene of Baikal environmentalism.

Called "Putin's favorite oligarch" (e.g., Reguly 2011), Oleg Deripaska began his career as a destitute student of theoretical physics in the Soviet Union, but, seizing new opportunities available in Russia in the 1990s, he used his financial acumen to accumulate a diverse portfolio of Russian industrial enterprises. While building his financial empire, Deripaska battled openly with other rising oligarchs during the decade of fraud, corruption, and mortal danger that followed "shock therapy" and privatization (e.g., Shelley 1995; Varese 1997; Frye 2002; Appel 2004; Shevtsova 2007). Although wealth inequality in Russia is among the highest in the developed world (World Bank 2012; Credit Suisse 2013), a minority of academics has begun to challenge the narrative surrounding Russia's oligarchs. Once vilified as virtual thieves who enriched themselves at the nation's expense, these new voices praise the oligarchs in contemporary Russia for their role in curbing the chaos of the 1990s and shepherding the country toward stability and profitability in the 2000s (e.g., Guryev and Rachinsky 2005; Gorodnichenko and Grygorenko 2008; Treisman 2010). Oleg Deripaska was a chief architect in Russia's transition from mafia-style capitalism to global economic power, and he personifies the ambivalent nature of the Russian oligarchy.

Deripaska intersects with environmental activism around Lake Baikal through his sudden and extensive involvement in projects of "corporate social responsibility" (cf. Wartick and Cochran 1985; Margolis and Walsh 2003; Carroll and Shabana 2009). Local environmental activists in Irkutsk must confront the meaning of this new corporate partnership in environmental preservation. Some were suspicious of En+'s motivations, while others worried about the potential threat to activist autonomy. One organization even chose to reject corporate funding. But, despite their ambivalence, local activists have overwhelmingly decided to accept En+ money and don its corporate logo. The involvement of En+ in environmental projects around Lake Baikal demonstrates the central role of branding in the way corporations siphon civil power from social organizations and draw it into the economic field. The recent arrival to Russia of a practice that slowly evolved in the West allows us to see more clearly the power relationships which undergird it.

The growth of corporate social responsibility through "cause marketing" is more than greenwashing and less than environmental stewardship. It is the means by which two distinct holders of generalizable power trade a little of what they have for that which they cannot themselves possess. Cause marketing is an attempt to pull the glamour of civil society into the economic realm and onto corporations—and in exchange, social organizations receive the financial power that a nonprofit cannot produce by itself. Thus, the field of power can explain a practice otherwise rife with contradiction, moving our understanding beyond debates over cooperation and co-optation.

August 14, 2012

The twenty-three members of our eclectic volunteer work crew hiked for an hour from the platform where the train had stopped, through the *taiga*, and finally we arrived at the work site. For the next two weeks, we would be living in this spot in the woods, building a switchback trail from the river valley to the ridgeline several hundred feet above. But before getting to work on our main project, we first had to build the base camp (Figs. 5.1 and 5.2). We split up into teams: two volunteers dug a pit to bury organic food waste; several worked constructing a large table that twenty people could sit around for meals; some people gathered fire wood; others built the rack that would suspend our cooking pots over the camp fire. When all these tasks were finished, we got to work setting up tents.

GBT provided tents for the volunteers, and we slept two people to a tent. These were nice, big, new tents, and we pitched them close together in the small clearing. The forest was soon home to a clump of green and yellow domes, like mutant mushrooms growing from the pine straw floor. Each tent had a small "foyer" made by the outstretched rain fly, and the door to the foyer bore the GBT logo in white. Beneath the logo, in extremely large script, was the word *Ent* (Fig. 5.3).

Figure 5.1 A GBT base camp with constructed tables, benches, and cooking fire. Photo Credit: Nancy Hollman

Figure 5.2 A GBT volunteer cooks dinner for the 23 member work crew over the open flames. Photo Credit: Author

Figure 5.3 A tent provided by GBT to its work-camp volunteers, emblazoned with its own logo and that of its corporate sponsor, the En+ Group. Photo Credit: Nancy Hollman

After the chores were done, we gathered in a circle. Katya passed out favors to thank us for participating. We all got matching blue bandanas that quickly became a collective badge of honor among the members, who donned them daily as we set out for the trail. The triangular bandanas were made of synthetic material and covered with small GBT logos printed in white. And alongside each GBT logo, written in large lettering, was that funny word *Ent* again. Clearly it was some kind of brand, but nothing I had encountered before.

In the evening, I was sitting around the campfire with a number of Russian volunteers and I ventured to ask: "What is Ent?" My question was met with confused faces and some shrugs.

"What is *what*?" one of them asked me in return. I repeated the word and then pointed at the logo on the tents.

"Oh, I know this!" one of them suddenly announced. "I know these people." She made a sour face. It was Polya, an environmental scientist. She corrected my error—what I had seen as the letter *t* was, in fact, a plus sign. The logo was that of a large corporation called the En+ Group. "En+ is an energy company and they build hydroelectric dams," she explained to me. "China wants to develop in its northern region and they need a lot of power. So En+ is talking about damming rivers in Russia in the Far East and sending the power to China. But who will profit from this? En+ will get all the money and China gets all the power. But what does Russia get? Nothing. And what does nature get?"

She ended there, as though the answer to this question were obvious.

"So, they are not good stewards of nature?" I asked.

"Not at all," she shot back. "They are only interested in money and profit. But they give money to organizations like GBT so they will look environmental even though they are not. They only care about money . . . They give a few tents and then drown rivers and build dams."

Such was my introduction to the En+ Group.

Oleg Deripaska and the En+ Group

En+ Group Ltd. is a holding company that operates mining, metals, energy, and logistics businesses throughout Russia and abroad.[3] Headquartered in Moscow, although registered for tax purposes in Jersey, En+ was founded in 2002 by Oleg Deripaska as a privately traded collection of assets in a wide range of natural resource extraction and energy production enterprises. The centerpiece of its assets is a controlling share of United Company RUSAL (Russian Aluminum), the result of an aluminum industry consolidation that Deripaska began in the early years of privatization. Through acquisitions and mergers, Deripaska and his partners have developed RUSAL into the largest aluminum producer in the world (Bell 2016). Additionally, En+ owns the coal company VostSibUgol, which mined 16.8 million tons of coal in 2011, most of which En+ consumes in its own factories and plants, and EuroSibEnergo, which builds and runs numerous hydroelectric dams and nuclear power plants. EuroSibEnergo is the largest independent power producer in all of Russia. Employing more than 100,000 people worldwide, En+ Group reported revenues in 2011 of US $15.3 billion.

To better represent the vast energy and material empire that exists under the auspices of En+, the following portrait comes from *BusinessWeek*'s investment summary of the privately traded company:

> En+ Group Ltd . . . produces alumina, aluminum, and bauxite; and smelts aluminum for consumers in defense, aviation, transportation, ship-building, packaging, and construction industries. The company also generates hydro and thermal power; provides engineering services to projects in the fields of power generation, grid infrastructure, housing facilities, metallurgy, and oil industry; distributes energy raw materials, including coal, gas, and oil to consumers of the wholesale electricity market; produces energy coal; and supplies solid fuels for thermal plants. In addition, En+ Group Ltd. produces ferromolybdenum, which is used in the manufacture of stainless and certain other steels, and heat-resistant nonferrous superalloys; and steel, oil, industrial machinery, chemical, defense, electrical, and electronic industries. Further, the company engages in the integrated process of ore mining; ore processing to produce molybdenum and copper concentrates; and molybdenum concentrate roasting for ferromolybdenum production through smelting. (BusinessWeek 2013)

The majority of En+'s plants and facilities are located in Eastern Siberia, which, although remote, is touted loudly as the company's greatest asset. "Close to fast-growing Asian markets" is the oft-repeated refrain in En+ promotional materials. The company has ambitious plans to develop and exploit the natural resources of Siberia, viewing the region as the linchpin of Russia's continuing economic growth.[4] En+ points to the geographic proximity of Siberia to China as a vital competitive advantage over mineral producers in Africa and South America.

> Asian nations already consume more than 30% of the global energy output, more than 50% of steel, [aluminum], copper and other metals, and about 2/3 of global iron ore. Asian consumption will only grow [in order] to fuel rapid industrialization and urbanization. New

Chinese cities alone have seen an inflow of 100 [million] people from rural areas over the past ten years, which has created a huge demand for housing, roads, infrastructure, and transportation system construction—which in turn requires more and more resources. A further 250 [million] people are expected to move to new Chinese cities by 2020—and a similar transition will be witnessed in India, [w]hich has a population of over 1 [billion], Indonesia, Vietnam and other emerging regions of the world. The Asian economies are going to develop most rapidly in the next few decades—and the demand for resources needed to maintain the economic growth rates will constantly grow. (En+ Group 2013)

En+ aims to meet this exponentially growing demand with the buried wealth of Eastern Siberia. As the company explains: "Eastern Siberia is blessed with some of the world's most abundant stocks of natural resources, where 90 percent of the Russian platinum group metals (PGM), about 70 percent of nickel, copper and other metals, 80 percent of coal reserves, substantial hydrocarbon reserves are stored . . . All these riches have been left almost intact, with the region being socially and economically underdeveloped" (En+ Group 2013). En+ intends to unleash the potential of Siberia's latent resources, piggybacking off Asia's emergence as a major player in the global economy. Siberia can be tapped to meet the growing consumer demand fostered by urbanization and capital penetration into China's interior. Such is the strategic development plan laid out by the President and founder of En+, Oleg Deripaska.

Oleg Deripaska[5] is among the richest men in Russia,[6] with a net worth of nearly US $9 billion. His life is a rags to riches story that could only have been possible at a particular juncture in Russian history—he was one of the few individuals to seize the moment in an era of mass privatization and a burgeoning capitalist economy.

In the late 1980s, Oleg Deripaska was a student, studying quantum statistics at Moscow State University in hopes of entering the field of theoretical physics. At the time, in the late Soviet period, there was still no financial market, but change was in the air. *Perestroika* had opened a Pandora's Box of Western ideas and practices long forbidden by the Communist Party. As censorship loosened its hold, debates about democracy and liberal economics found their way into the newspapers and magazines on shelves throughout the country (cf. Yurchak 2006). Deripaska says that he became interested in investing after reading an article about securities exchanges in the late 1980s. After the formation of the Moscow commodities exchange, Deripaska bought a seat, where—as he says—he began applying the principles of physics to calculate the buying, selling, and trading of commodities, in association with his partners.

Russia in the 1990s was a country buffeted by crisis and chaos. The transition from the state-planned economy to the market was not the slow, controlled phase-in that reformers in the vein of Gorbachev might have imagined. Instead, the country faced "shock therapy" and the immediate privatization of state-owned enterprises.[7] Without the Soviet state to strong-arm production or police the population, companies ground to a standstill (Kotkin 2008). With a mounting debt crisis and hyperinflation, the bankrupt state left millions of workers' wages in arrears (Desai 2001). The entire Russian economy was privatized virtually overnight through a process that has been widely disparaged as corrupt and without sufficient regulatory oversight (Hoff and Stiglitz 2004). Most Russians gained nothing from privatization, while a few who were well connected or nefarious walked away rich (Appel 2004). Managers and owners of the newly privatized firms began an epidemic of asset stripping. Pyramid schemes erupted, and gangs of mobsters started terrorizing the streets for "protection money" (Varese 2004). Corrupt dealings, fraud, insider trading, rigged auctions: all of these and more characterized the decade following the collapse of the Soviet Union.

The young investor Deripaska was not immune to this history; his meteoric rise in the 1990s could not have been accomplished without shaking dirty hands. He was only twenty-six when he took control of his first aluminum smelter in 1994 (Kramer 2006). He continued to consolidate his hold on the aluminum sector throughout the 1990s, during a period that is known as the "aluminum wars." Throughout the decade, there were multiple contenders who fought for control of Russia's aluminum industry, its fourth largest export, and the fighting was not in the variety of clean competition. Battles were often brutal, crime-ridden, and characterized by financial fraud and physical violence, which cost many individuals their lives. As oligarch Roman Abramovich describes it, "someone was murdered every three days" (Peck 2011). By the end of the "aluminum wars," more than one hundred people were dead, including industry managers, bankers, traders, politicians, and mafia men (Ahmed 2012).

Deripaska's individual role in the dirtier side of Russian business appears ambivalent from the outside. He has been the subject of multiple lawsuits alleging nefarious business dealings (Kramer 2006). He has been investigated by several countries for money laundering and links to organized crime (Perez and White 2009). In 2001, he was uninvited from the World Economic Forum conference in Davos (Bershidsky, Rozhkova, and Tru 2001), and in 2006 he was embarrassingly denied entry to the United States (Simpson and Schmidt 2008). While Deripaska has never been charged with a crime and strenuously denies any wrongdoing, as one journalist wrote, "it will probably take years to dissipate the stench of criminality surrounding the industry he now controls"

(Klebnikov 2001). Deripaska admits to paying organized crime for protection in the lawless decade of total institutional collapse after the fall of communism (Ahmed 2012). In this, he claims he was not unique; it was the only option for those doing business within Russia in the 1990s. Times have changed, business in Russia is much cleaner, and Deripaska played a leading role in that transition.

In 2000, Oleg Deripaska sat down at a table with rival oligarchs Lev Chernoi, Iskander Makhmudov, and Roman Abramovich. At the invitation of Abramovich, the group hammered out a merger of their aluminum companies, signaling an end to the violence and a new regime for the commanding heights of the Russian economy (Kochan 2003, Gardham 2012). Their various competing regional interests were joined into RUSAL, a single company, under the management of Deripaska, that would collectively control 70 percent of all aluminum production in Russia. No longer would the captains of industry battle each other in internecine struggle; instead they would unite for the creation of a national industry that would springboard them to a wider, global marketplace. The meeting has since come to symbolize the end of *dikii* [wild] capitalism in Russia with its mafia-style business dealings, and a shift toward an established, stable, national corporate monopoly.

On January 1, 2000, the same year that Deripaska helped end the aluminum wars, Vladimir Putin ascended to the presidency. His rise similarly indicated a shift in the field of power. Putin reasserted state control and refilled government coffers, and in doing so he also established new terms between the Kremlin and its oligarchs. Namely, Putin made it clear that he would not tolerate oligarchs meddling in politics. Those who did would soon feel the full force of state power. Media tycoons Boris Berezovsky and Vladimir Gusinsky lost their assets to the state and went into exile when their criticism of Putin garnered his ire (Gessen 2012; Harding 2013). Oil magnate Mikhail Khodorkovsky lost his giant Yukos oil company to the state and found himself imprisoned for ten years on charges of fraud and tax evasion (BBC 2013b). With little incentive to augment their power through political involvement, the economic elites turned toward perfecting the power that already defined them. They set their sights on the global stage.

The 1990s were socially, politically, and economically catastrophic for Russia, but the new millennium brought about a decided shift in orientation. The opportunities of the global market helped discipline a chaotic economy, and Putin's enforced separation of economic and political power promised the country, if not rule of law, then order at the least. In light of these changes, a small number of scholars have expressed a more positive view of Russia's oligarchs, showing the high productivity and efficiency of their industries when compared to those retained or managed by the state. Treisman (2010) has shown that the companies that were auctioned in the "loans for shares" deal have since significantly outperformed those that remained in state control. Gorodnichenko and Grygorenko (2008) found in Ukraine that oligarchs were more likely to invest in productivity-enhancing upgrades to their facilities and to vertically integrate their production, resulting in greater profit and efficiency. Rather than greedy gangsters, the new outlook on the oligarchs suggests that of the reform-minded "citizen employers" (Haydu 2008), responsible for bringing Russia into its present era of stabilization and progress.

Oleg Deripaska personifies this new conception of Russia's "responsible oligarch." Staunchly apolitical but avowedly patriotic, Deripaska states that he is interested in building business rather than simply making money. Under his control, the Sayanogorsk aluminum smelter increased its output even beyond Soviet levels. However, when aluminum faced a global crisis in overproduction and a significant drop in its commodity price, Deripaska became the darling of global investors by agreeing to keep his factories operating below capacity (Williams 2013). Deripaska is active in the World Economic Forum at Davos, and has been a longtime crusader for a more transparent, rationally organized, Western-style business community in the Russian Federation. Among his adopted practices is the phenomenon of management stock options, as an incentive to align company leadership with investment outcomes, à la American-style "shareholder capitalism" (Gordon 1999; Dore, Lazonick, and O'Sullivan 1999). In 2013, Deripaska received much positive press by foregoing his $3 million bonus and using the money to purchase $25,000 in company stock for 120 of his 72,000 employees (BBC 2013a).

Moreover, he has been involved in extensive charity work and public service. Deripaska, along with several of his peers, has come to regard the question of his legacy as one of great importance. He sits on several boards of trustees, including the Bolshoi Theatre, the School of Economics at Moscow State University, and the School of Business Administration at St. Petersburg State University. He has also donated more than $8.5 billion to charities and social causes through his private foundation, Volnoe Delo. Deripaska, along with several other major Russian oligarchs, has been aggressively seeking a more positive image for himself and his companies abroad. In the end, some say, posterity will measure these tycoons not on how their fortunes were created but on how they shaped the country.

En+ as Corporate Philanthropist

AUGUST 21, 2012

Back in the *taiga*, half way through our work trip, my GBT crew had again gathered in a circle before heading off for the trail.

"Before we get back to work," Katya began with a mischievous smile, "I have a little surprise gift for you!" She opened up the potato sack she was holding and pulled out a blue tee shirt.

"Clean clothes!" exclaimed one of the Americans in the group, and people laughed, since clean clothing was truly a novelty after a week in the woods. Katya showed off the new shirt. It had the En+ logo on the left breast. The back of the shirt bore the GBT emblem. We each came forward and claimed a shirt in our size.

"What's this?" Yegor asked, pointing to the En+ logo. He looked at the back and the front, flipping it over twice. "It's like advertising," he announced and then laughed loudly.

En+ became a major player in the environmental scene in Irkutsk in the fall of 2011. Representatives for the corporation's new sustainable development division showed up in Irkutsk and sought out all the local environmental non-profit organizations with a single message: we want to be your corporate sponsor. By the next year, Irkutsk was awash in environmental activities sponsored by En+. Some of these were projects the company initiated; others were conducted in partnership with local nonprofits. In the case of GBT, En+ simply offered money for material supplies to support the projects GBT was already conducting.

In late January, I traveled to Moscow and visited the corporate headquarters for the En+ Group. I was there to interview a representative from the En+ sustainable development division, which manages the company's corporate social responsibility work. The building was tucked away on a small, winding street, close to several national consulates. From the outside, the headquarters building was fairly nondescript. Even the interior, at first pass, seemed uninviting. There was only a tiny foyer with tightly packed and uncomfortable seating, fluorescent lighting, and a dirty white ambience that reminded me more of a clinic than a corporate office. The desk personnel, a young man and a young woman, struggled mightily to speak to me in English, even though I had greeted them in Russian. They were expecting me, they said, and haltingly tried to explain that I had to wait. When I assured them they could speak Russian, a look of gratitude swept over their faces. I was given a visitor's pass that I could swipe at the turnstile to allow entry, but only after my host came to greet me.

Soon another woman, a secretary, came to collect me and usher me upstairs. Again, I was unimpressed by my surroundings until we left the stairwell and entered through a main door. Suddenly the ambience shifted; I was surrounded by what has become the standard décor for a corporate office. Gone were the uncomfortable clinic chairs and dirty white walls. I was directed to a small conference room and seated at an oblong table. I came to think of this as the "Green Room:" key-lime walls, Kermit-green chairs, forest-green industrial carpet. A large potted tree was growing at the far end of the room. My escort gave me a set of En+ promotional materials and told me that my interviewee was running behind schedule. She then ordered me tea, which was brought to me on a silver tray with two chocolates in wrapping emblazoned with "En+."

After about ten minutes, a representative of En+'s sustainable development department arrived. I asked her to tell me about her programs and, evincing great pride, she launched into a detailed description of the many activities that take place under the auspices of En+'s sustainable development division.

The entry of En+ into Irkutsk environmentalism was as sudden as its money was abundant. The general motivation for the company's philanthropy is two-pronged: first, social responsibility has become the established norm for multinational corporations; second, environmental philanthropy helps to counteract the stewardship balance sheet for a company with a less-than-stellar environmental track record.

When I asked the representative about the creation of the En+ sustainable development division, she immediately pointed to Deripaska's study of business, particularly international business, and his longtime participation in the World Economic Forum at Davos.

EN+ REPRESENTATIVE: Deripaska . . . is a person who learns things for himself and participates in a variety of foreign programs—he has studied in America and in Switzerland, he has various degrees—he is always learning. [His studies on] international management systems gave him the idea that he should split his businesses into two groups. One group would be those businesses that must steadily develop, grow, and provide growth to the country. The other would be investment businesses that he would buy, develop, and sell, in order to make money. The first business group is . . . En+, and it includes those companies that he will always own, that will continue to develop for a long time and will develop Russia as a country. It is heavy industry . . . The company was eventually ready to start developing on social issues and sustainable development, because it should eventually do this. Business can't start on day one, or maybe they can but not in Russia . . . Therefore, it is just in this moment, a couple years ago, we came to such a level of development ourselves that we were ready to begin implementing such a complex and encompassing social project. So we created this separate division.

The global influence on corporations such as En+ was confirmed by another informant I spoke with in Irkutsk who is close to both local nonprofits and the business community.

> It's quite a change for big business. Russian big business is still very young; it's literally only about 20–25 years old. In such a short space of time it's hard to become a civilized business. And so Russian business usually does its own thing. [But] when a business enters the international market, it discovers that [international business] has its own rules of the game and certain standard requirements to ensure that you and I can negotiate on equal terms. Simply put, you have to meet certain criteria, including how socially responsible you are, how you act in those areas where you do business, how much assistance is provided to charity, and so on and so forth. And our business understands that it must be done. Not because you want to, or don't want to, but because that is what is expected. And since it is expected, now we'll do it.

Essentially, En+ did not invent the model it has adopted for its corporate giving, nor was it instigated by conscientious members of the company. Instead, it was a new division, with orders from the top, that was created in the image of Davos.

En+ Group's interest in environmental partnership also comes, admittedly, out of the company's past and present sins against nature. The corporation and its CEO have been at the embarrassing center of a number of environmental fiascos. Most notably, in the early 2000s, Deripaska's other holding company, Basic Element, acquired the Baikalsk Pulp and Paper Mill, which has been a source of popular ire since it was first constructed in the 1960s. In the 2000s, local environmentalists again built a major campaign to close the factory. In 2008, a court determined that the plant was breaking the federal Law of Baikal and had to cease production until it could operate with a closed wastewater system (Alexandrova 2013). The plant stopped operation. Initially, the company refused to pay wages to its furloughed employees, resulting in massive labor unrest with workers, who also pointed the finger at Deripaska. Putin intervened, staging a media opportunity where he played the great patriarch dressing down the oligarch, who hung his head in public shame (Barry 2009). Deripaska then offered a personal loan of 150 million rubles to cover worker compensation (TASS 2012). Meanwhile, Deripaska pleaded with Putin, claiming that the legally required environmental upgrades would render the mill unprofitable. In 2010, Putin overturned the court order, granting an exception to the Baikalsk mill. The government allowed BTsBK to continue production as long as it instituted the closed wastewater system by 2012 (Harding 2010). The decision to reopen the mill without environmental upgrades created a new wave of environmental protest, and Deripaska once again became public enemy number one. He sold the mill in 2010 (Russian Financial Control Monitor 2010).

The representative of corporate giving for En+ repeatedly strove to disassociate Deripaska from the infamous paper mill. During our discussion, the En+ representative described their work on "mono-cities," the Russian term for a "company town," and she brought up Baikalsk as an example. Then she added that it was a bad example, since En+ does not own the mill. Somewhat disingenuously, she went on:

> A long time ago, like ten years ago, Deripaska bought it, and then sold it. I mean, he didn't build it. But people continue to have the sense and the belief that it is his business. And because of this, Deripaska's negative image here has remained. Although he only controlled it personally for two to three years—not very long. It hasn't been his for ten years. But because the city is in our region of influence we decided to include it also.

At the same time that Deripaska was dealing with the fallout of the paper mill, En+ was the target of another environmentalist campaign, this time on the detrimental effects of the massive Boguchansky Dam on the Angara River. This colossal hydroelectric dam was originally planned and designed under the Soviet government, but it was not constructed due to lack of funding. In 2006, En+ joined with state-owned RusHydro to complete the dam. One of the largest dams in the world, Boguchansky would power a massive aluminum smelter that En+ planned to construct nearby. Environmentalists were adamantly opposed to resuming work on the dam. They claimed it was poorly designed, with few of the modern safeguards normally required for maintaining a healthy river system. They also claimed that its construction was nonsensical because the dam was located far from any population center with no infrastructure for transmitting the power it generated. While En+ plans to construct a large aluminum smelter, the plant does not yet exist as of 2017, and neither is there a working population to support it. Finally, whole villages of indigenous Siberians were displaced by the flooded reservoir with hectares of uncleared forest submerged. The Boguchansky Dam is now complete and running, but only after years of vocal opposition from local and national environmentalists.

These are the two most notorious examples of environmental harm resulting from industries under Deripaska's control, but the very acts of mining, metal processing, and conventional energy production may also harm the environment to greater and lesser degrees. To this extent, En+ will always have a conflict of interest with environmentalist organizations. True to the model of corporate social responsibility, the company is attempting to bring environmentalists in as community stakeholders while simultaneously giving to environmental causes to bolster their public persona as environmental stewards.

The sustainable development representative for En+ was straightforward in her explanation of this fact in responding to my question as to why they were giving to environmental causes in Irkutsk.

EN+ REPRESENTATIVE: Large industrial enterprises always affect the environment. Always. [Whether they are] good, bad, American, Russian, European, from a certain point of view, any industrial undertaking is harmful to the environment, despite the fact that there are economic advantages. Therefore, in order to develop more appropriately, we have to make sure these industries take into account environmental factors, risks and dangers. So, in environmentalism, we can say that there is a technical side, like, say, the environmentalism in the business, factory modernization, installing various filters. That is the technical part of the enterprise, which is done there [in the factory]. And there is a second part, which is global, and has to do with the issue of environmental development of the region. This is what is included in our social program . . . To say it rudely, industry inflicts the most damage on the environment in any situation, whether it is good or bad. That means that, in making money, [industry] absolutely should try to fix those problems that are bringing harm. And a conscientious company understands that. A non-conscientious company will try to fight with environmentalists using various means. We try not to fight, but to find some common ground and partnerships to address these environmental challenges, whether our production is guilty or not.

Unable to avoid the tarnish of environmental culpability, En+ has developed a proactive approach that seeks out relationships with environmentalist organizations. They offer support and engage in partnerships, meeting the needs of environmentalist organizations and seeking input on their own industrial development. This new approach to relations between industry and environmentalism could be said to be cooperative or co-optive, and whether it is one, the other, or possibly both, can be in the eye of the beholder. Environmentalists in Irkutsk looked on at this corporate stance with great ambivalence.

Environmentalists React to En+

It was my day to cook for the GBT work crew, and I slipped into the cook shack to chop vegetables. When I came in, I realized that I was not alone. Polya and Katya were already in the shack, seated on the wooden benches behind the table drinking tea. Polya was questioning Katya about GBT's relationship with En+, something that seemed to bother her deeply.

"We talked about it for a long time," Katya explained. "The club discussed it, because we know they are a big corporation and we didn't want them to control us. But in the end, we decided that taking the money wasn't giving them control. We are still independent. We still decide what projects we are going to do. No one is telling us what to do."

"But it isn't about whether En+ will control you," Polya objected. "It's that it is so little money for them and they get so much for it."

Katya did not register Polya's comment that En+ is actually gaining something from its donation to GBT, and instead fixated on the size of the donation compared to the corporate budget as a whole.

"Yes, it is only a little money to them, but for GBT it is enormous!" Katya said

"But they make that money hurting the environment," Polya countered.

"And we talked about that," Katya said.

But we decided that we could use this money for good. Why not take money that came from bad and turn it toward doing good? And we are not the only ones. Defend Baikal Together—which is another nonprofit that does environmental education in schools—they get money from En+ and now they have a new monthly bus tour that brings children to Baikal who might not otherwise go. They could live their whole lives two hours from Baikal and not know about it. And it isn't just En+! Lots of nonprofits take money from corporations. When I worked for Earth Corps in Seattle, they took money from big corporations, like Boeing and Microsoft. And they also discussed whether they would lose their independence, whether the corporations were using them, but they decided that taking the money didn't mean the corporations were using them, because they were going to do the work anyway, only now they had better equipment.

She started to list the things that we had on our trip that had been paid for by En+: pick-axes, tents, shovels, tee shirts. Polya still seemed unsatisfied.

"Baikal Wave might have a different perspective," Polya ventured to suggest.

"Yes," Katya agreed. "They likely would have a different perspective. They probably wouldn't take money from En+."

"And En+ would not give it to the Wave!" Polya retorted.

Polya's assessment proved to be incorrect: En+ did seek out Baikal Environmental Wave for a partnership, but they were met with a cold shoulder. Jennie Sutton mentioned the encounter briefly on my first day as a volunteer with Baikal Environmental Wave. The organization had recently become involved with One Percent for the Planet, a movement to connect nonprofits with companies committed to donating one percent of their profits to environmental causes. Sutton stressed that the Wave would not take money from just any company that wants to give.

"We are very selective with whom we work," she told me. "Very particular. We require that there be some environmental component to the business. For example, there was one company that really wanted to work with us and support us. They have their sights set on some big damming projects on some rivers in Siberia and selling the energy to China. They would like to get us, an environmental organization, under their thumb, but that is *not* going to happen!"

Later, in another context, Jennie Sutton and Marina Rikhvanova were discussing Oleg Deripaska. While I now knew En+ from the logo that was plastered on so many environmental projects in the region, I was still unfamiliar with the man behind the curtain.

"Who is Deripaska?" I asked.

"He's an oligarch who owns lots of various energy and mining companies," Jennie said. "His company tried to come here and sponsor us. What were they called? Something-plus? It doesn't matter, it's all Deripaska."

Baikal Environmental Wave was the only organization in Irkutsk to reject En+'s offer of partnership. They were also the only nonprofit that virtually never referred to the corporation by name. In the Wave's office, the companies under the control of En+ or Basic Element were called by their metonym: Deripaska. He became the personification—the active and culpable agent—for environmental harm committed by the companies he owned. The Wave's apparent refusal to adopt the corporate moniker seemed intentional, as though the brand were a smokescreen to hide the puppeteer. While the En+ name and logo found its way onto many environmental projects throughout the region, in the Wave's office, it was a non-entity; they only discussed the oligarch Deripaska, naked without his brand.

The Wave was unique among environmentalists in the region in their firm stance not to associate with En+. Most others, while not conceding their right to criticize the company, were nonetheless glad for the money and the partnership. Elena Tvorogova from the organization Reviving Siberian Land hoped that the interaction would not only help her organization, but might even spur new ways of thinking in the minds of Russia's elites.

ELENA TVOROGOVA: En+ appeared at the end of the year before last, toured around, met with practically all of our environmental organizations. And to each organization they proposed that they were ready to support any of our projects, so what projects do you have? Obviously, every organization then had to make a decision: Do we wish to enter into such a relationship with such a business? For example, Baikal Wave decided no, we are not going to play these games. [But Reviving Siberian Land] has been working with business for a long time, already for many years. We have done projects with money from Yukos and money from Polyus Gold, and now with money from En+. Because I think it's better to do a little bit of good work with money from business, than not to do any work. Because business will spend the money somewhere anyway. They will find somewhere to spend it. But they could spend it on something stupid. So I think it is much better to use this opportunity for good . . . I don't want to overstate our influence on the minds of our oligarchs. Still, some of these business representatives, experts working in this field, coming into contact with these themes and problems, you will see them, just a bit, a little bit, maybe 2 to 3 percent, turn Green.

At GBT, members were ambivalent about the relationship, but finally came down on the side that it was better to accept the money than to not. There was only one member—Lusiya—who was adamantly opposed. The sponsorship proceeded over her objections, and she had to make her peace with it. But in an interview, she explained her concern.

LUSIYA: When Lena asked us about this last year, it was in the fall I think, she just said that it is possible to get such a grant, and she asked "What do we think of this?" I was the only person who was strongly opposed. Very much against it. And I'll explain why. I said that an organization, especially if it is a social organization, it must be consistent in its actions. And I thought that it is inconsistent when we, say, go to a protest for preserving Lake Baikal and demand that something is not built, that they do not reopen the paper mill owned by [the CEO of]

En+, and, I mean, I think it is just unethical. You need to differentiate whose money to accept. But since the question was asked to all, and most people agreed, the decision was taken. And [pause] I still think it's such a double standard. Because, on the one hand, we are an environmental organization, and to take money from an organization that pollutes the environment—that's bad. On the other hand, I think it's better if the money goes to something good, than if they spend it on something else. Because at least I'm sure that the money that is given to GBT will be well spent. I know that it will be for educational projects, I know that it will be for the trails. I know it will be helpful to a variety of environmental initiatives. So, let En+ at least do something good. I mean, this is their chance to do something good. And . . . well, I cannot say that I am completely "for" it, but at least I think that it's, you know, as a lesser evil [laughs]. But in any case, I think that GBT is compromising its conscience.

However, Lusiya was singular in her unilateral opposition to En+. More often, comments from GBT members resembled those offered by Pavel. He described the debate over whether to accept their sponsorship as a "hot topic," but fundamentally decided that he was in favor of the relationship.

PAVEL: I know all about [En+] and what they do. And I don't hold deep, positive feelings for them, like, "Ah, how good! What a great company!" I understand that for them it is just PR, I guess. Because they are trying to be like a Western company, right? But I think it's good, and you shouldn't reject these companies simply because it's PR for them. Let all these companies do it for the purpose of public relations, but they are giving money for good activities . . . I think it's also payment for how they harm nature, and pretty strongly at that . . . But! Here's another additional point. That said, I believe that if they give money, then that does not mean that they can affect our activities. Or tell us what to do or not to do. Or that we can no longer talk about them using unfriendly words. So that am I supposed to say, "En+ Group is a wonderful company." No. That is, I believe that it is not. There are certain concrete decisions that I have not liked. For example, tee shirts. I believe that this tee shirt is for GBT, not for En+. But for some reason I see En+ and I don't see GBT [on the front]. That is, if I had the chance, I would not myself accept this shirt because that's not what I need. I think this is the wrong approach, precisely in this concrete issue. And the same thing with the signage, when they made badges, emblems where you can't see GBT, but there

is En+, I wouldn't have agreed to that. I mean, I think it's just a little
bit too much.

The ubiquity of the En+ logo made it an easy target for environmentalists to
suggest that the company was only interested in its image. But logo battles
reflected a greater concern among participants that En+ would threaten the
autonomy of GBT. The fear of corporate domination would creep up in GBT
as random shows of autonomy that were arguably overblown.

One such example was GBT's ten-year anniversary, which was held in
December of 2012. GBT had incorporated as a nonprofit in 2002, and the
anniversary was seen as a real triumph and reason to celebrate. Few nonprofits in Russia reach ten years, I was repeatedly told, and the collective wanted
to plan a self-congratulatory bash. However, big celebrations take discretion-
ary funds that volunteer organizations seldom possess. Lena Chubakova, the
executive director, went to En+ to ask for money for a party. En+ immediately
agreed and told her not to worry, that they would take care of everything: the
emcee, the program, the venue, and so forth.

"I told them, no, we are a nonprofit organization and we are completely
capable of throwing our own anniversary party," she told the club meeting one
night. "Because what if they invited all their own people and not ours? What if
we ended up with the En+ logo all over everything and nothing about GBT?"

The same refrain was repeated for the next two weeks as the group worked
on the details of the event. When discussing the program, Lena again said,
"We have two options: 1) we have En+'s designers put together the program,
which I'm not too keen on, because who knows if they will put a giant En+
logo over everything; or 2) we do it ourselves." Lena was certainly thrilled
to receive En+'s financial support, both for the party and for the many other
aspects of GBT's work. But her fixation on the omnipresent En+ logo made
clear her underlying fear that En+'s self-promotion would come at least in part
at the expense of GBT's autonomy. Her fierce stance on the anniversary party,
in itself not a dire issue, displays a territoriality on the part of an organization
that fears being subsumed by its wealthy benefactor.

Clearly, environmentalists were mostly concerned with three issues in their
interactions with En+: 1) they were concerned that their autonomy would be
undermined and En+ might control their activities; 2) they were worried that
En+ was only doing it for the sake of self-promotion; and 3) they were worried
about their own reputations, or perhaps at a deeper level their organizational
souls, being sullied by association with a large, profitable corporation that has
been known to harm the environment.

All of the local organizations except for Baikal Environmental Wave were able to reconcile themselves to these three concerns. First, while the blatant self-promotion of corporate giving was met with rolled eyes, there was general agreement that it is better for corporations to make an effort toward having a social conscience than to do nothing at all. Second, they were generally satisfied that they retained their autonomy; indeed, many claimed that the added resources only enhanced their ability to achieve what they intended to do anyway. Finally, environmentalists in Irkutsk came to the conclusion that rather than tainting themselves by association, they were instead doing greater service by helping to turn ill-gotten gains toward a socially and environmentally beneficial end.

But there is a deeper logic that undergirds the rise of corporate social responsibility, one that exceeds the often interminable debates about corporate co-optation and the beneficence of business. It is a logic that springs from the field of power. Once we recognize these two groups as the bearers of different types of power, the relationship can be seen as transactional: trading one power for another. And just as in any such exchange, there are consequences for the choices that are made.

Cause Marketing: The Cost and the Benefit

Corporate social responsibility is a business ontology. The "presentation of self" (Goffman 1959) that an enterprise adopts in relation to this ontology is known in the trade as "cause marketing" (cf. Varadarajan and Menon 1988; Smith and Alcorn 1991; Earle 2002). The principle behind cause marketing is that a for-profit and nonprofit organization can enter into a partnership for mutual benefit: the nonprofit receives financial aid and the company garners positive attention to its brand. Cause marketing is more than corporate philanthropy, which is described in the literature as a donation to a nonprofit and whose benefit to a for-profit company lies principally in the tax deduction. The goal in cause marketing is broader: cause marketing aims to create an associative link in the public mind between a company and a good cause. The return to the company is in image, reputation, and public goodwill, as well as increased sales. Studies have shown that cause marketing is both cheaper and more lucrative than traditional marketing campaigns (e.g., Smith and Alcorn 1991; Bloom, Hoeffler, Keller, and Meza 2006; Krishna and Rajan 2009). A cursory glance through journals of marketing and business will show the great effort that has been put toward research that aims to maximize returns to a company in cause marketing: assessing the relative impact in choosing one's cause, or the type of support (e.g., Ellen, Mohr, and Webb 2000; Nan and Heo 2007). Essentially, cause marketing is an investment, upon which corporations expect to see a profitable return.

Despite the general recognition that En+ was motivated to fund Irkutsk's environmental scene as a public relations coup, there was little acknowledgment, or perhaps even understanding, by my informants that sponsorship can be viewed as a market transaction. For the activists, the money is still viewed as a donation; the activists and En+ are operating on different planes and from widely divergent perspectives. For the environmental activists, the money donated is contrasted to their income *before* the sponsorship. From their perspective, the question is one of *money versus no money*. This "either-or" approach to sponsorship helps explain why the environmental activists focus principally on whether or not to associate with En+ as a partner, and why they do not discuss the amount they receive. When they do discuss the size of the corporate sponsorship, it is usually in relation to the organization's regular income; from this perspective, Katya could say that GBT is receiving an "enormous" sum from En+.

Yet for En+, the transaction is different and amounts matter. By its association with myriad environmental causes, En+ is investing in its image. The company is essentially purchasing the right to associate itself with various environmental groups. When associating with a nonprofit like GBT, the question for En+ is not *money versus no money*, but rather what is the price of GBT's reputation? En+ receives increased name recognition, and has its brand associated with a positive, feel-good cause, both directly, through volunteers, and indirectly through the mass media. The affiliation raises the positive profile of En+ locally in Irkutsk, nationally—given the importance of Baikal—and even internationally through GBT's global volunteer base. From this perspective, affiliation with GBT is an enormous bargain. For the price of a tent, they reach countless people sleeping in those tents, trip after trip, year after year. For the cost of a silk-screen logo imprinted on a blue tee shirt, En+ imprints its brand on the minds of multitudes, from pedestrians passing volunteers on the street to those viewing GBT trip pictures on the organizational website. Environmentalists in Irkutsk see their work as a cause; En+ sees it as a commodity—and one it is acquiring at a terrific discount.

This same interaction can then be re-examined as an exchange by different players in the field of power, each of whom commands a different and distinct arsenal according to that player's societal position. Social organizations cannot produce money; to the extent that they can utilize that generalizable power, it must come secondhand. Similarly, when corporations align with nonprofits to enhance their reputations, what they actually desire is civil power: the ability to exert influence by means of one's virtue (see Chapter 1). But because these institutions operate for private profit, they cannot inherently command civil power. It is beyond easy reach of their particular societal position. Instead, agents of capitalist corporations have learned to trade power for power, and garner the glamour of virtue by standing in the halo of their nonprofit partners.

However, despite the apparent contradiction between the civil power that arises through selfless work for the common good and the private profit that forms the basis of business enterprise, some corporations still try to "cut out the middle man" and attempt to build civil power "in-house." By directly engaging in social service projects, these corporations seek to direct public goodwill directly to themselves *as though they were* social service organizations. Since these businesses are, in fact, doing work toward the common good, they are likely successful in accruing some civil power to themselves. But when it comes to actual social outcomes, there is good reason to suspect that it is better to trade power with a nonprofit partner, whose principal mission is service, than to try to forge a different arsenal de novo.

When corporations do attempt to build civil power through their own "good works," there arises a conflict of interest: it is only civil power that is desired, and this power is more readily achieved through "presentation of self" than through the long, arduous process of actually producing socially esteemed results. In the case of environmentalism, when the image of stewardship becomes the primary point of an activity, the cause itself is liable to suffer. The quest for a positive headline or an eye-catching photo opportunity can result in activities whose cost may not equal the benefit environmentally. To this extent, environmentalists in Irkutsk are correct in suggesting that it is better for them to accept corporate funds for their own projects rather than to allow En+ to spend its money on "something stupid." In fact, En+ does conduct its own environmentalist projects in addition to those it sponsors with local organizations, and the emphasis on publicity leads to decidedly ambivalent environmental outcomes.

360 Minutes for Baikal

Litter has become a major concern around Lake Baikal. As the region has developed into a major domestic outdoor tourist destination, the traffic to Baikal has increased. Yet the infrastructure around much of the lake remains primitive. Most of the territory surrounding the lake is protected—including two territories with the highly restricted *zapovednik* status.

Where villages and outposts dot the lake, there is no trash collection. Residents are supposed to haul their trash to a landfill, but since many lack transport, they also frequently burn it or bury it in the woods. Such a transgression by villagers would not amount to vast environmental damage except for the advent of two phenomena: tourism and globalized mass consumer production. The amount of packaging waste arriving into the villages has grown exponentially since the Soviet era; now, with the influx of tourists, this trash has only increased.

"Packing out what you brought in" has not yet become standard outdoors etiquette for Russian tourists. Vacationers will boat up to the lake shore; picnic or camp; and leave their beer bottles, tin cans, and plastic bags behind them. Some people will even dump their broken lawn chairs or tents. The more conscientious will dig a hole for their rubbish in an improvised, and unregulated, mini-landfill, but these accumulate and expand with remarkable rapidity, leaving many of Baikal's beaches sullied. Litter, dropped into the water by boaters, will also wash up on the shore. In the Soviet era, the forests may have been able to absorb what little was left behind. Today, with the plastic packaging of mass consumer production, and with the greater number of visitors, the sheer amount of waste renders these practices unsustainable.

Since litter has become a visible blight on Baikal's shores, it is also an easy target for environmentalist concern. Many list it just behind the paper mill as the chief threat to Baikal. Even those who are not active environmentalists generally recognize the problem of accumulating litter. Lately, a number of initiatives have arisen to collect rubbish. Schools in the region regularly organize clean-up projects. In the city of Irkutsk, a coalition of nonprofits, state agencies, and businesses called "We Will Make Irkutsk Eco-Logical" [*Sdelaem Irkutsk Eko-logichnii*] coordinated a massive work day to rid the city of illegal dumps and litter.

En+ joined the litter collection effort with its own event "360 Minutes for Baikal," an annual clean-up project that takes place in association with Baikal Day. At its inaugural event in 2011, one hundred volunteers traveled to the shore of Baikal to gather trash. I attended the 2012 event as a participant observer. I had been recruited by members of GBT. En+ had enlisted the help of GBT members as crew leaders, who would each direct small teams of volunteers. Although it was not their own project, the GBTers were glad to lend their enthusiasm and expertise.

I had been in Irkutsk for only a month when I was recruited to 360 Minutes for Baikal. The summer was nearly over, and many people were glad for an excuse for one last trip to Baikal while the weather was still warm. I was encouraged to go at least in part as an opportunity to visit Olkhon Island, where the litter clean-up was to take place. Olkhon is the largest island in Baikal and is also a major tourist destination. The island, with its unique topography of rock formations, ice caves, forest, and steppe, as well as its endemic flora and fauna, has long been sacred to the indigenous Buryat shamans. Dina, one member of GBT, said she had signed up to participate specifically so she could visit Olkhon, a place she had never been. I had also heard of Olkhon and was eager to see it; helping keep it clean and observing local environmentalists in action would be an added bonus.

SEPTEMBER 9, 2012

I awoke before sunrise to go, as I then thought, to Olkhon to pick up litter. There was a surprising number of people congregating on the corner of Kirov Square, and lots of taxi-vans and buses to take the volunteers to their destinations. In all, En+ had recruited four hundred people to volunteer for the day— a fourfold increase over the previous year. I quickly found the GBT folks in the crowd, and there were happy hugs all around.

"Where are you going?" Vasya asked me.

"To Olkhon," I answered.

"Ah, nice," he replied. "We are going to a campground on the mainland, but across from Olkhon, just about fifteen kilometers away."

Anatoly from GBT was one of the crew leaders. He came up to me and said, "You are in a group with Yanna and Pavel. Stay with them, and you will be in the right place. Your group is standing over there." He indicated a small cluster of people nearby.

"That's my group," said Vasya.

"But if I am in your group, then I'm not going to Olkhon either!" I realized out loud.

I was not the only one disappointed. Dina from GBT came up to me later and said, "Remember how I told you that I had signed up for this because I had never been to Olkhon in all my life? Well, it turns out I'm *still* not going to have been to Olkhon. They have me in a group that is on the mainland." I wondered whether the increase in participation was not partly related to the mistaken impression of a trip to Olkhon.

Our crew leader was a young man named Vova. He had to check us in before loading us on the bus. We were the last group on the bus, but there were still plenty of seats. We sat and waited. Then a young woman named Tamara, who was the leader of another crew on our bus, announced: "If anyone needs to use the bathroom, you need to go now because you won't get another chance for two hours." I asked where we could go to use the toilet, but Tamara didn't know. Pretty soon people were pulling out their cell phones, calling around to see if local businesses were open yet so we could use the restroom. Pavel turned to me and said, "That building there is Irkutsk Energy." He indicated the formidable cement structure, whose door was flanked by two security guards, next to which the buses were parked. "It is owned by En+, the company that is organizing this action, and yet they won't let us inside to use the toilets." I was confused and mildly irritated that someone would say "You should use the toilet now," when there was not a viable toilet available. Eventually Irkutsk Energy relented and opened its doors to volunteers who needed the restroom.

Finally we boarded the bus and began our long, arduous journey. At first, it was a pleasant morning drive through the country, watching the world turn from city to forest, and forest to steppe. Little grey marmots kept popping up and watching us drive past in our caravan of buses.

On the way, the crew leaders addressed their teams and gave out the schedule. The trip from Irkutsk to the work site should take four hours. We would eat breakfast at a campground at 11 a.m., spend six hours (360 minutes) collecting rubbish, return to the campground for supper, and then leave on the buses again, arriving back at Kirov Square around 10 p.m. However, according to the schedule, we were already forty minutes late in departing from Kirov Square.

The crew leaders also passed out baseball caps and windbreakers to the volunteers, each bearing the En+ logo and an emblem for the project: "360 Minut Radi Baikal" [360 Minutes for Baikal].

"There will be news media and television cameras filming, so you must wear the windbreakers and hats at all times," Tamara told us. "Do not take them off."

"What if I get hot?" someone asked.

"You cannot take them off," Tamara repeated. "Maybe take off what you are wearing underneath."

Now for a word about the bus we were on: it was a Korean bus, decorated with drapes and tassels in the windows. Ornaments in the shape of the Playboy bunny hung for inscrutable reasons from every other window. We were part of a caravan of six buses, with a police car in the lead and in the rear. I was unsure of the purpose served by the police car because it certainly was not helping us reach our destination more rapidly. The trip seemed interminable. First, the paved road turned to dirt. Then the dirt road turned to bumpy rock. And eventually, the road disappeared entirely when our caravan was forced to take an off-road detour. The bus had virtually no shocks, so in addition to a bumpy ride, we also were traveling at a snail's pace. According to the schedule it should have taken no more than four hours to reach the work site, but four hours came and went, and still we were driving.

At five and a half hours into the trip, I leaned forward to Yanna and Pavel, who were sitting ahead of me.

"What do you think, are we going to stop sometime?" I asked them.

"Yes, sometime," they both responded.

"The bus will run out of gas eventually," Yanna added.

She was right. After more than six hours of travel, we did eventually stop at a campground called Danko. We disembarked and were ushered to the cafeteria. According to the schedule, we were supposed to have "breakfast" at 11 a.m. Now, it was 1 p.m., and we were sitting down to rice kasha with salami and bread. One of the GBT volunteers complained later about the bad food: "Those tiny slices of bread that were, as my grandmother would say, cut like in a restaurant!"

Outside the cafeteria, there was a porch set up like a stage. Large banners of bright orange bearing the En+ logo waved in the breeze. All the volunteers were asked to gather on the stone steps as makeshift risers above the improvised stage. The media were instructed to set up around this stage in a semicircle (Fig. 5.4). There were multiple television cameras and several reporters from print media. I found myself standing beside a woman in her thirties from Krasnoyarsk, another city in Eastern Siberia. She introduced herself as a journalist. She was wearing the obligatory hat and windbreaker, and I learned I was not the only "participant observer" there.

"En+ did a similar project in our city," she said, "so I am here as a follow-up story." I asked what she thought of the project. "It is fine and all to pick up the trash. But it would be better if people didn't litter in the first place."

After a staged event where representatives from En+ thanked participants and partners, talked about safety, and made short speeches for the cameras,

Figure 5.4 The improvised stage for the photo opportunity/media event in conjunction with the 360 Minutes for Baikal litter clean-up event. Photo Credit: Author

we were sent off for the day's mission. We were all given orange drawstring bags with logos on the outside and lunch on the inside: bottled water, three *pirozhki*, an apple, and two granola bars. Each work crew also received gloves, trash bags, shovels, rakes, and a wooden sign that read, "Please don't litter." Then, we piled into yet another vehicle to be driven still further away to our respective work sites. Fortunately, my team's work site was only about five minutes from camp.

Despite the stress of the day, it was impossible to miss the beauty of the Olkhonskii *raion* [county]. The campground where we gathered was one of several tourist bases on the Western shore of Baikal, in an area named Maloye More, or "Small Sea," after the shallow channel between Olkhon Island and the mainland. Baikal water is warmer here, and it has become a popular spot for tourism. The landscape was full of fascinating shapes; the treeless steppe accentuated the contours of its undulating topography. There were strangely shaped rock islands sprouting up from the water amid the many peninsulas and bays. There were horses grazing by the bank where the van dropped us off.

We uncovered several trash pits, mostly filled with glass bottles and burned cans (Fig. 5.5). One volunteer spent a great deal of time fishing disintegrating plastic out of the water, joking sardonically that it was a bag from the previous year's litter clean-up. He also pulled up what looked to be motor oil. Someone left an old raft and a broken lawn chair by one camp site. The trash truck came by early and we were able to get some of the bigger stuff hauled away. Undoubtedly, the foulest things we uncovered were several used baby diapers. From then on, no matter what we were pulling out of a trash pile, we would say, "At least there are no diapers!"

While we worked, picking up glass, plastic bottles, and metal cans, I asked about recycling and why there was no recycling in Irkutsk. I heard someone say there used to be recycling, and I asked why it had stopped.

"They had recycling in the Soviet Union," Pavel explained. "But the thing was that, back then, there just were not that many products. Not like now. You maybe had three kinds of drinks, and they were all made locally, or at least regionally. So back then, the companies would take the bottles back and you would get a deposit. Not a lot of money, but something. Now, bottles are coming from far away, from all over, and it is just cheaper and easier for companies to make new bottles than to collect and reuse the old ones. But we've never had the kind of recycling where you could just bring any glass or plastic or metal. It was just those specific bottles going back to that particular company."

We kept collecting trash until we ran out of plastic bags. At the end, we planted the wooden sign on which En+ kindly asked visitors not to litter. The work was over and the silliness ensued. Two boys started swinging bags of

Figure 5.5 360 Minutes for Baikal volunteers clean up an improvised landfill. Photo Credit: Author

trash at each other. Yanna came up behind me and pretended to brush my hair with a rake. We took some group photos and then all clambered back in the van to return to the campground. We left the bags of trash piled up and were told a truck would come collect them. There were some looks of skepticism exchanged, but there was nothing for us to do but hope that they would, in fact, be collected.

When we arrived back at the campground, volunteers were being stopped by a videographer from En+ wanting to take footage for an in-house promotional video. The volunteers were asked to say their names, where they were from, and why they were there. Some of the television stations were also continuing to collect interviews with volunteers. We returned to the cafeteria for supper: soup, salad, and pirog. After supper, we piled back in the bus for our long journey home.

On the road back to Irkutsk, I talked briefly to Pavel, a geographer working on his PhD. He was examining the effect of large businesses like En+ working in remote regions, such as Bolshoye Goloustnoye or Olkhon. I asked what his results were and he made a face.

"It's mixed. More negative than positive, though. The thing is, they may do some little things that are good, but usually all of the local wealth will be sent out of the region. Big businesses do not actually create good jobs for local people, just bad, service kinds of jobs."

Despite the many problems, though, he was pleased that large businesses were making an effort at doing something positive. "Like this event," he said. "It's all about PR, but still, five years ago this kind of thing would never have happened. There were no kinds of volunteer opportunities and no one picking up trash five years ago."

Trash on Baikal is a real concern. But "360 Minutes for Baikal" did far more to benefit the company that created it than it did the lake it was ostensibly supposed to serve. Even the name was something of a misnomer; because of the many delays, volunteers were only actively gathering trash for about two hours. The effect on the lake was minimal; there was less litter, but only by a very small percentage. And yet the positive publicity that this event generated for En+ was clearly substantial. The story reached local, national, and international (English language) media outlets. Footage and photographs clearly displayed the En+ brand on banners, backpacks, and especially on the windbreakers and hats that volunteers were forbidden to remove. The story was a positive one of a company making an effort to clean up a beloved national treasure.

En+ was clearly willing to pay a great deal of money for the publicity. The company hired buses, vans, and drivers to haul hundreds of people from Irkutsk to the Olkhonskii *raion*, and then paid for hundreds of volunteers, staff, and media guests to eat two hot meals at a campground and a picnic lunch. It provided apparel and souvenirs to all participants, not to mention the necessary tools and equipment, the police escort, and the vehicles to haul away the collected rubbish. Had the principle concern of En+ been the state of Baikal, one might imagine many more cost-effective means to achieve comparable, or even superior, results.

Given the amount of auto transport involved—twelve hours of driving, with multiple vehicles that were not full to capacity—it is arguable that the environmental footprint of the event exceeded the benefit of the litter collection, a fact not lost on the environmentalists involved. In particular, the members of Great Baikal Trail bemoaned the cost-benefit of a one-day trip. Their organization, with its "volunteer vacations," also has a practice of carting volunteers to distant locales for environmentally beneficial projects. But the environmental cost is weighed against two full weeks of labor. In the case of 360 Minutes for Baikal, participants spent six times as much time in transit as they did volunteering. When I asked later what he thought of the event, one GBT member said, "For all the money they spent on it, you would think they would have let us spend the night and actually get some work done!"

But En+ was not paying for the work. They were paying for the publicity, and that could be accomplished just as easily with two hours of work as two days, and the former would not involve overnight accommodations for hundreds of volunteers. In a publicity stunt, the size of the event is more important than the quality of the experience. Had trash collection been the major motivation, En+ could have revamped the annual event to that end. However, when I discussed the project with the representative from sustainable development, she said that the plan was to expand the project so that volunteers were cleaning trash all around the circumference of Lake Baikal. And, indeed, the next year, the number of volunteers more than doubled to a thousand, and the trash collection took place at sites all around the lake—an increase in quantity, but not necessarily in quality.

The En+ representative admitted in our discussion that education, not trash collection, was really the goal. "If kids, especially teens, spend one day picking up trash and see how hard it is, then they will no longer litter. And they will tell their friends and family not to litter . . . The goal is that eventually [the event] won't be necessary." She also hoped that participants would be introduced to volunteering and would continue to do so.[8]

These two goals are in accordance with the environmental movement that the company aims to serve. Environmentalists like Jennie Sutton would dismiss local litter clean-ups as "feel-good" measures that produce no lasting structural change. But most of the local environmentalists were glad that litter clean-ups were taking place. As one pointed out, in order for structural change to take place: "First, people have to *see* the litter." En+ aims to help the public to see the litter—as long as they, and the mass media, see the En+ brand prominently alongside its removal.

Conclusion

Corporate capitalist business has only been a player in the field of power inside Russia for a brief twenty-five years. In the first decade following the Soviet collapse and the introduction of private enterprise, the minority who succeeded in amassing fortunes from the remnants of Soviet industry spent their days in bitter and sometimes fatal internecine battles to dominate the economy. Consumed by the struggle in their own social field, they did not attend to the greater field of power. The Russian state was itself in shambles and posed no serious threat to oligarchs or activists alike (although both suffered the negative effects of state structural collapse). Flush with foreign financing, Russian civil society had no strong incentive to turn to domestic business for assistance, and Russian capitalists seldom saw the benefit of allying with social organizations, who were themselves equally new and whose role in society was still murky at best. While Oleg Deripaska fought the "aluminum wars," he was not concerned with what environmentalists were doing. It was only with the emergence of the Russian oligarchs in 2000 as a unified class with shared goals and objectives that their relationship with civil society became more complex.

As oligarchs like Deripaska set their sights on the global market, the globalized field of power seeped through them into domestic business practice, including cooperation with local civil society. As Russian industrial magnates moved into a more prominent position within the global economic market, they found themselves beholden to previously established norms of global business. Among these was an insistence upon corporate social responsibility (e.g., Bartley 2007). While late in coming to the social responsibility mandate, companies such as En+ have since moved full throttle forward, seizing the opportunity to align with nonprofit partners. For environmental activists in Irkutsk, affiliation with a corporate partner amounts to a windfall, opening new opportunities that would have been otherwise beyond their reach. But to corporations such as En+, the relationship is expected to have long-term economic payoffs. The expense is minimal, and the rewards are potentially vast.

The tactic of trading financial power for civil power was not invented by En+ and the Baikal environmentalists; it was learned in the course of other struggles in the field of power between civil society and multinational corporations in the United States and Western Europe in the late twentieth century. The new skill of trading power was then widely shared in global social networks, such as the World Economic Forum, and only recently brought to Russia, where the meaning of corporate social responsibility and the practice of cause marketing needed to be examined and explained locally as new and highly ambivalent phenomena. That the practice is common in the West helped grant it legitimacy in the eyes of local activists, some of whom still looked upon a relationship with corporations in disgust.

To understand corporate sponsorship as a trade of power between players in the field of power is to expose the logic that undergirds many of the contradictions within the practice. Why do environmental organizations cooperate with businesses that have unsavory stewardship records? Why does the public face of corporate giving often overwhelm the work that actually takes place? These two societal groupings are each able to wield generalizable power, but their power source comes from a position that renders their access to the other's power source problematic. Consequently, they engage in a trade through cause marketing.

Cause marketing is not greenwashing, whereby a company engaged in environmental harm seeks to advertise better stewardship than it actually conducts. Neither is it pure philanthropy, as would be a simple tax-deductible donation. Instead, cause marketing is how corporations purchase access to civil power. That power may be used to better the image of the company for the sake of higher profits, either in recruiting new customers or in garnering political goodwill in order to forestall regulation. Companies use their purchased civil power in their own social fields, as competitive advantage in burnishing their brands. The emphasis on branding may be so great that the "cause" itself becomes window dressing, as in the case of 360 Minutes for Baikal, where the environmental impact of the clean-up project was arguably less than positive.

The trading of power for power does not necessarily upset the balance within the field of power as a whole. In fact, the practice has the curious effect of maintaining a certain level of proportionality among the players involved. In order for cause marketing to work, social organizations must retain their civil power; they must possess it in order to trade it. They have to continue to embody the virtue that gives them strength: working selflessly and voluntarily for the common good. Cause marketing actually *demands* third-sector autonomy for its success: without a "virtuous" civil society, there can be no halo for the corporation to bask in.

However, an autonomous and powerful civil society can still threaten economic elites in the broader field of power. Although capitalist corporations need a strong and independent civil society for successful cause marketing, and they even serve to increase nonprofit capacity through their financial contributions, these same corporations simultaneously confront their nonprofit partners as potentially hostile opponents in the field of power at large. It is a precarious position they must negotiate, and they turn to power plays in the larger field of power to manage the tension.

6

Disempowering Empowerment

Chapter 5 shows how civil society organizations trade the power that accrues to its good deeds for the power of money generated by the for-profit sector. Meanwhile, businesses that operate for their own self-interest need an independent civil society to bask in the glamour of its virtuous voluntarism. But the "worthiness" that gives civil society its power is a double-edged sword for business elites. For cause marketing to be effective, civil society must be independent and honorable. But an independent civil society, buoyed by the strength of conviction, holds civil power, and that power can be used against business interests. Civil society can make demands upon business that constrain the aggregation and deployment of financial power: they may target corporations directly through boycotts, awareness campaigns, and the promotion of alternative products. Civil society can also act indirectly by calling for government regulations, new restrictive laws, or higher taxes. Corporations need an independent civil society, but they also confront it as a potential opponent in the field of power. Corporations and business elites have a strong incentive, therefore, to limit civil power while preserving its independence. They do so through the same partnerships that provide them the benefits of cause marketing. Sponsors can choose programs that work to promote business values, while also preaching environmental protection. By selecting the types of environmentalist program to fund, businesses can protect their practices from environmentalist critique. In the process, they may work to limit the power of civil society and successfully alter the balance of the field.

This chapter discusses how corporate sponsorship works to employ civil actors in the propagation of a pro-business style of environmentalism. As it embarks upon its program of sustainable development, En+ chooses particular programs to fund, and eschews others. Doing so, En+ uses civil society to spread its own message. Environmentalists find themselves championing the market and promoting business logics along with the cause they hold dear. When this happens, environmentalists inadvertently constrain their own source of power; if business values are considered morally good, then environmentalist critique of business lowers the reach of its civil power. When En+ uses its money to fund environmentalist projects that hold business values as equal to or higher than environmental ones, it is making a play in the field of power, constraining civil society's legitimate claims and shoring up its own power position in the field.

In making such a power play, En+ is not inventive or unique; it is reproducing domestically a relationship that has already become dominant in many parts of the world. The particular case documented here at Lake Baikal mirrors the so-called fourth wave of environmentalism[1] in the West (Dowie 1996), also promoted by the theory of ecological modernization (Buttel 2000; Huber 2000; Young 2001), both of which seek to harness the power of markets for environmental purposes. The strategy of fourth-wave environmentalism is for activists to prevent business backlash toward their demands by cooperating with business needs. However, "command-and-control" regulation is often a more effective means to pollution abatement than market-based measures or voluntary corporate action.[2]

The ideas of fourth-wave environmentalism were not imposed upon Russia; instead, this global trend paralleled a homegrown distrust of government and the collective memory of a command economy that was also none too kind to the environment. Rather than debating the relative efficacy of required versus voluntary measures, what the Russian case brings to the fore is the means by which cooperation between business and environmental civil society can work to insulate business from environmentalist critique, irrespective of any particular problem. En+ preempts environmental critique not by disavowing its claims but by fostering social projects that place environmental concerns below the normal needs of business in the public mind. Importantly, En+ does not make this claim itself, which could be contested as self-interested. Instead, it utilizes its relationship with environmental nonprofits so that environmentalists themselves promote business logics in their own work.

Disempowerment by Design

Civil society organizations are frequently engaged in socializing projects. "Socialization" is the name given to the process by which symbolic systems are acquired and accepted by a public (e.g., Hyman 1959; Merelman 1969; McFarland and Thomas 2006). Educational programs are often the primary mechanisms for socialization. Public school systems in the United States assimilated immigrants, socializing them into a common American culture (Graham 2005). Youth subcultures socialize schoolchildren of various economic classes toward appropriate workplace dispositions (Willis 1981).

Even efforts aimed to counteract the discriminatory impact of socialization through "empowerment" programs cannot totally transcend the dominant culture: teaching young people to navigate the society around them necessarily replicates the terms of that same society. When business gains access to the operations of civil society through corporate sponsorship, it acquires the ability to put civil power to work for economic ends within the field of power.

THE SCHOOL FOR ENVIRONMENTAL ENTREPRENEURSHIP

The School for Environmental Entrepreneurship, or SHEPR, as it is called according to its Russian acronym, is a recent initiative launched as a partnership between En+ and a local nonprofit organization called Reviving Siberian Land.[3] SHEPR is a boot camp for budding entrepreneurs. Admission is via application, and while the majority of participants are in high school or university, there are also some working adults in the program, mostly in their twenties or thirties. Over the course of five days, participants are involved in nearly nonstop activity, geared toward developing and refining the business plan for an environmental enterprise. Students may apply with a project in mind, or they can apply independently and join another student's project after admission. The school takes its participants and their business proposals through a series of intensive learning and evaluating mechanisms, culminating in the presentation of a final business plan before a jury of potential investors. What begins as a very loose, undeveloped idea grows into a defensible, implementable, entrepreneurial venture.

The school also functions as a competition. Each team seeks to garner the most points by participating in and successfully fulfilling a variety of events and benchmarks. At the end of the course, points are tallied and a grand winner is announced. Winning the competition has very real, as well as symbolic, rewards; the winning team leader is awarded 4,000 rubles to further the project and move it toward implementation.

The inaugural session of SHEPR was held in August 2012. I attended the second session in February 2013 as a participant observer. Several occurrences from my experience illustrate how the values of business repeatedly trumped environmental stewardship, despite the intention of environmentalists involved in the project.

Day 0
Despite its name, SHEPR largely presumes knowledge of environmental issues and green business. No aspect of the program actually teaches environmental practices or mindsets. The only such instruction occurred the night before the

program actually began. Not all students were yet present, and no points were offered for attending the presentation. It took place at the end of a busy day, filled with travel and orientation. But, despite these drawbacks, the activity clearly had an impact on those who attended and participated. The presentation involved a lecture and educational game by Zinaida, a volunteer from Baikal Environmental Wave. A professor of pedagogy in her sixties, Zinaida had designed and developed the game, which she called "Eco-Footprint of Commodities" [*Eko-sled Tovary*[4]].

Zinaida began by laying a hand-drawn map of the world on the floor. "There are about seven billion people in the world" she told us. "So let's have seven volunteers get up and stand on this map. You can each represent a billion people." Several students obliged. "But of course you can't stand everywhere on the map. You don't have a billion people in the ocean." The students bunched up on the land masses. "So everything that comes from the Earth, all the land, the trees, the farms, the forests—this is all we have for all these seven billion people. There is nowhere else they can go. Nowhere else we can get resources from. How do you feel?" She asked.

"Pretty crowded," someone replied, and there was mild laughter.

"Look at your feet on the map. Think of this as your environmental footprint. You are taking up a certain amount of the Earth's resources. So this is how it would be if we all shared the Earth's resources evenly among the seven billion people. But that isn't the case, is it? Some of us take up more than others. Some of us have bigger footprints." Here she retrieved a pair of giant "shoes" made out of cardboard that were covered with magazine clippings of advertisements and products. She handed them to one of the students.

"Here put these on," she instructed, and the young man did so. "Now what happens?" The young man with the shoes took up most of Eurasia and Oceania, and some of the others started to tip over and fall off the map.

"We have less," said one of the girls astride North America.

"There are fewer resources for everyone else," Zinaida confirmed. "But what if it were equal? What if all of you had shoes like that?"

"It would be impossible," several people said.

This led to a lecture and slide show about resource use and waste that was pointedly critical of consumption. It was after 9 p.m., and it was clear that not all students were equally present and alert, but a cluster of them were engaged, and I occasionally heard a comment or a reply to a question that showed that critical connections were being made.

Zinaida then held aloft a Kinder Egg. This is a common children's treat in Russia: a hollow chocolate egg wrapped in foil with a plastic toy inside. The chocolate is enriched with extra milk and vitamins, so it is considered healthier than other candies, but to kids it is still chocolate.

"You all know what this is," Zinaida began. "It is likely a part of everyone's childhood. Maybe you once collected the prizes inside. Now we will take this one product and look at its environmental footprint. What are the resources that it takes to make this one little egg?"

Here she pulled out a box and set it on the table. The box was filled with cards showing pictures of either raw materials such as water, wood, sun, or animals, or processing centers such as mines and factories. Some cards had forms of transportation, like trucks, trains, planes, or boats. Students clustered around the table. They were instructed to take every component of the egg—the chocolate, the foil wrapping, the plastic toy, and the paper instructions—and create four threads that traced the piece from raw materials to finished product. They needed to think of where the rubber came from, how it got to Russia, what went into chocolate, how many stages it took make the foil, how many times products had to travel, and how much electricity was needed at each stage. At the end, students could see the long chains they had created, and how much is necessary for that piece of candy, sitting as an "impulse buy" beside the cash register of most grocery stores. There was only one table and one set of cards, so only about eight of the forty people in the room were actively interacting in building the eco-footprint. The attention of those in the back waned early. But the exercise seemed to impact a few of the more participatory students, who looked at their work and said, "Wow!" or "I had no idea!"

After the game wrapped up, we returned to our seats and Sasha, SHEPR's second-in-command, stood up to tell us what to expect tomorrow. But before discussing logistics, Sasha encouraged students to think about the "amazing egg."

"Isn't it incredible how many things go in to making this one tiny egg? Businesses all over the world contribute to it. Lots of enterprises play a role. There are famers, truckers, factory workers, and your business can be just one part of that whole." What had been a lesson in unsustainable consumption was shifted into praise of supply chain economics. Zinaida, who had spent the last hour conducting students through the game, stood nearby, tight-lipped.

Day 1
Today the school began in earnest. Registration tables were set up in the restaurant's foyer. After paying our extremely modest participation fee of 250 rubles ($8.75), we received spiral-bound program materials and acquired a preponderance of En+ branded goodies. Then we gathered in the restaurant that had been rented out to host SHEPR. There was a projector screen set up before several rows of plastic red and blue stools. Elena Tvorogova, the director of Reviving Siberian Land, stood and gave the opening remarks to a seated crowd of students before her.

"Welcome to the winter session of the School for Environmental Entrepreneurship!" she cried, and we all applauded. She went on to describe the school and what we could expect to gain from it.

"Only those projects that receive a certain number of points and can attract enough people to the team will get to move forward in the competition. If your idea turns out not to be successful, that isn't failure—that is *experience*. Your job, then, is to think up an even better idea based upon what you've learned. This is what you have to do in the real world as an entrepreneur. So your business idea doesn't work? Think of another idea."

Elena Tvorogova continually emphasized that the school would mimic the "real world" as much as possible, particularly in the final activity. "At the end of the school," she said, "you will stand and present your idea before a panel of investors. These are *real* investors from Sberbank and the Small Business Administration. They are not used to gushy, emotional language. They think: quick, precise, and serious. This is not a game. This is a real conversation with serious people. They will determine who wins the competition."

Next she introduced Sasha, the second-in-command, who continued to introduce the program.

"The world is a big system, a big network, and you can take advantage of it," Sasha said. "Learn from each other. When mental barriers start to fall, that is great for starting a business. You can use each other in this way to get new blood. You should be like vampires—the Twilight vampires—good, kind vampires." We laughed appreciatively. "You want to bite lots of people and get more blood. This is called social capital, so capitalize!"

He then went on to say how fortunate we were to be able to participate in such a program. "This is just like the business development sessions that you can find in other places, but those cost ten times more." At SHEPR, En+ would foot the bill.

There were several talks scheduled before we could begin working on our own business proposals. One of these was given by a representative from En+ named Ksenia.

"What is sustainable development?" she asked us with a big smile. "What do you think when I say 'sustainable development?'"

Students began to call out possible definitions: "When you don't lose any money," one said. "Working within the law," answered another. "Monopoly," another added. "It's about stability," guessed another. "The quality of your products, and your workforce," was the last guess.

Ksenia smiled again. "I am going to tell you what En+ is doing for sustainable development and hopefully when I am done explaining, you can have a better answer for the question: What is sustainable development?"

"Whether small or large, a business's image is what guarantees its success long-term," she told us. "One's image or reputation is extremely important.

You can spend years building it up, and in a single moment it can be destroyed. When we talk about image we are talking about it with a wide lens. We are talking, not just about your product, but how you show yourself in the territory where you work, whether that is in a small village or worldwide.

"Many companies, especially new ones, think only about profits. But 30–40 percent of the success of a business is due to its reputation," she continued. "Twenty to 25 percent of the stock market price of a company is due to its reputation. Sometimes that number can even be as high as 80 percent. Think of reputation as the price of the materials that go into a product relative to the price of the brand. For Coca-Cola, the ratio is 4:96—4 percent materials and 96 percent brand.

"So let us apply this to the sphere of social responsibility. Your business is spending money when it works on social responsibility, but you *should* spend money on it. This is an investment in your image." Ksenia then went into detail about the various programs and projects that En+ supports in Siberia, including SHEPR. "Anyone can earn a profit," she concluded, "but you have to work hard to have a positive image and a good reputation."

After this talk, we moved on to the main event of the day: the project fair. Students spent the two hours before lunch preparing posters about their projects that they would display at the fair. Students and staff were given stickers to affix to posters of projects they liked. Stickers represented "points" and only those projects garnering a certain number of points could advance into the competition. Any team with fewer than five members also would not make it to the next round, and soon my roommate Veronika and her brother Pyotr recruited me to their team. We were then joined by two more individuals who did not have projects. So now we were a team of five, and had collected sufficient stickers. Our projected, called "Eco-Field," passed into the competition.

"Clarify for me what 'Eco-Field' is all about," I asked Veronika and Pyotr, as we sat down at the table that we claimed as our workspace.

"So the idea is that we would employ about 150 people to go into an area of the *taiga* to live in tents and gather raw materials," Pyotr explained.

There is an herb that people use to make tea that grows wild in the *taiga*. People would live in the woods for four months, cutting and drying this herb. Individuals can gather up to five kilograms for personal use, but we could get a lease or agreement to take more. And then we would sell these herbs to a wholesaler in Angarsk. Right now there is a deficit of raw materials in forest products and the price is really high. These are natural forest products, and brands, like Taiga Products, Evalar, and Shalfei, sell natural teas and pine nuts. So there is a market for it. In the summer, we gather this herb, and in the winter we can gather pine nuts and *sera* [a sap that can make natural chewing gum]. Maybe we can also gather mushrooms and berries—all kinds of forest products. My dad runs a business like this near Olkhon. So I've done this before. My plan is to start my own operation near Baikalsk, only bigger. I already have the contacts at the wholesaler in Angarsk. I know the raw materials that they want and how much they pay per kilogram.

We were soon joined at our workspace by Marina Rikhvanova from Baikal Environmental Wave. She instructed us on the next activity, which was called: "Dreamer, Critic, Realist."

"So, when it comes to your business plan, what are your dreams?" Marina Rikhvanova asked.

"To be a famous Russian brand," Veronika said with confidence. "So that when people think of getting a Russian souvenir, they want to buy our product."

"Now for the Critic," Marina asked. "What are the critiques?"

There was silence. Veronika shrugged lightly and said, "I don't have any," then flashed a proud smile.

"I do," I said, and the group's attention turned toward me. "So you go into the *taiga* and collect your herbs. And you sell them and your brand becomes famous." Veronika smiled at this acknowledgment of her dream. "Now more people want your products, so you collect more of this stuff. And people love it, they go wild, they want more, so what do you do?"

"So we get more, and then more!" Pyotr said, egging me on.

"Yes! That is what we want!" said Veronika excitedly.

"This isn't a critique, this is a dream," Pyotr said, smiling.

"What I want to know is whether you have a mechanism to make sure you don't take too much."

They responded to me with confused silence.

"When you gather everything from one area, you go somewhere else," Pyotr answered.

"But when do you stop? How can you guarantee you won't harm the environment?" I asked.

"Because you don't kill the bush," Pyotr said. "This herb grows really fast. If you cut it all the way to the base, in four years it will be big and bushy again like before. So you clear everything in an area, but if you come back four years later, you can't even tell that you cut anything!"

"Okay," I said. "That's great that it grows back, but how can you guarantee you aren't harming nature? How do you know you aren't taking too much?"

"You can never take too much," Pyotr replied. "Siberia is endless!"

This word must have gotten Marina Rikhvanova's attention. She entered the debate, saying, "Nothing is endless. People always think that natural resources are endless, but everything has an end."

"And even if you aren't using up *all* the tea in *all* of Siberia," I went on, "you might decimate it in a particular region."

"But we don't kill the bush," Veronika explained to me with a patient smile.

"Okay, the bush will come back, but what if there were an insect that lives on that particular bush and now it has nowhere to live in that whole region? What if there is a type of bird that lives off that insect that is no longer there, and the bird can't eat? I'm not saying that this *will* happen, I'm just asking how you are going to make sure that it *doesn't*." Veronika's smile faded.

"Good brands for environmental products usually have some way of showing that they do not negatively impact the Earth," Marina said, cleverly steering the conversation back to branding and their concern to have a "good" business plan.

Veronika looked at me with awe. "I never thought about it like that. I never would have thought about birds." She looked as though a little LED light bulb were flickering over her head. Its illumination never reached her brother, however. Over the course of the school, Pyotr continued to insist to me that it was impossible to deplete the *taiga*.

Day 2

We began the day as usual, perched on our plastic stools at the back of the restaurant. Elena Tvorogova stood before us like a drill sergeant. She was dressed in all grey with a fleece jumper and reindeer boots. The silver pendant around her neck resembled a stop-watch, adding to the overall effect. She was our coach, our trainer. Throughout the day, she told us, we would be attending a series of master classes on topics essential to entrepreneurship.

First, I attended a workshop on "Creativity" [*Kreativnost'*] that introduced us to examples from the Disney corporation, de Bono's "Seven Hats" process, Howard Gardner's multiple intelligences, and crowdsourcing [*kraud-sorsing*]. My next workshop was on "Personnel." It regaled us with a laundry-list of methods for recruitment and selection, provided interview questions, and reminded us of ethical and unethical hiring practices. Finally, I attended a workshop on "Business Models." The instructor for that workshop made the only explicit allusion to environmental entrepreneurship that I heard all day.

"Seeing the whole system is especially important when you are doing eco-business," she reminded us, "because there can be environmental damage at any stage of the business process or at any of the source points." The moment stood out in its singularity.

Day 3

The chief event of the third day was the "expert conveyor." Each team had to go before a series of experts and defend its project according to a particular measure. There were twelve expert stations with the following themes: 1) attracting new people to your team, 2) media kits, 3) website design, 4) attracting resources to your project, 5) the case, 6) the break-even point, 7) competence profile and plan of development, 8) risks and their minimization, 9) marketing, 10) location; 11) eco-certification, and 12) preparing your presentation. Each station awarded up to three points. Teams were required to defend their project before at least six of the twelve stations to successfully complete the conveyor, although the more stations completed, the more points the team could receive. Consequently, most teams strove to complete the entire conveyor, and Eco-Field was no exception.

At the station for eco-certification, Pyotr sat across from Marina Rikhvanova and explained that the resources they gather from the *taiga* replenish themselves in four years.

"What will you do to ensure that this project won't damage the ecosystem?"

"We will get an ecologist to certify our project," he told her. "We will pay an expert to assess the territory, tell us how much we can collect, and give us their certification."

"What does that mean, a certification?" Marina asked. "How will that protect the ecosystem?"

Pyotr looked confused. "The specialist will know that the ecosystem is being protected. We will consult with experts who know and can provide certification."

"Certifications can be bought. You can pay someone; get a stamp on a piece of paper. But I want to know how you will ensure that this project won't damage the ecosystem." Pyotr looked helpless.

"Here is what I will do," Marina said at last. "I am going to give you one point." She marked a "1" on the score sheet and Pyotr's face fell and darkened. "I can't give you any more than that. I am worried that you go to the *taiga* and all you see is your raw materials. You don't see the system and how it all fits together. *You* need to be the expert. You are the one responsible, so you have to know *more* than the experts."

At the end of the day, all the projects were listed with the points they received at each station. Of all projects, Eco-Field had the second lowest score for eco-certification.

Day 4

Today was the culmination of all our hard work—the Big Day—presentations before the final jury. In the morning, the restaurant was still in disorder. People

with disheveled hair and dark circles under their eyes were frantically final-
izing their slides or practicing their presentations before their fellow group
members.

At noon, the final jury was brought in and introduced. These were five new
faces, individuals who were unfamiliar with our projects and how they had
developed over the course of the school. For that reason, it was said, they could
be expected to provide an unbiased assessment of our business proposals.
The jury included local investment bankers, a member of the Small Business
Administration, and the director of public relations from En+.

Each team took turns making a presentation before the jury. Our team
presented last. Generally, our performance proceeded as the others had, but
one moment stood out. The eighth slide showed the project's budget. All the
other projects had anticipated annual net profits that ranged from 500,000 to
1.5 million rubles, but Eco-Field's estimated net profit read 5 million rubles
annually. When this slide appeared on the screen, all five judges leaned forward intently, almost in unison, physically reacting to a profit margin that was
five times that of the other contestants.

We took a recess while votes were tallied. Marina Rikhvanova and I found
each other, and she asked how I had enjoyed the school.

"It was very interesting," I said. "But I am curious why there are no environmentalists on the jury." She looked surprised, as though this had not occurred
to her.

"I wonder why?" she asked aloud, and peered over at the jury for a moment.
There was a pause, and then she answered herself. "We ask uncomfortable
questions." She gave me a tight-lipped smile.

We all gathered back in the restaurant to learn the final results. Elena
Tvorogova gave a small, congratulatory speech.

"The future of the planet depends upon you," she reminded us. "No investor can tell you what you can do. Whatever you decide to do with your project
after the school, keep believing in it. Put your heart, your soul, and your time
into it."

And then the results were announced. First, they displayed the results from
the previous night. A bike trail project was in the lead, followed by a souvenir
shop, and then Eco-Field. Once the jury points were added, though, Eco-Field
handily won first prize.

ENVIRONMENT AND ENTREPRENEURSHIP IN SHEPR

SHEPR is built upon the fourth wave of environmental activism that emphasizes the creation of a "green economy" (Dowie 1996; Buttel 2000; Huber 2000;
Young 2001). However, there remains some tension and deep ambivalence in

the global environmentalist community about the implications of the fourth wave" In the halls of the academy and among activists themselves, there are those who proclaim that capitalism is simply antithetical to environmentalist aims (e.g., Schnaiberg 1980; Foster 2002; Speth 2008; Klein 2014). Others contend that without joining forces with big business, modern environmentalism is as good as dead (Nordhaus and Shellenberger 2007).

However, the School for Environmental Entrepreneurship suggests that there should be more subtlety brought to bear on the question of "greening capitalism." The question should not be *whether* but rather *how*. Fundamentally, SHEPR offers much more "entrepreneurship" than it does "environmentalism." Despite its name, and despite the involvement of committed and knowledgeable environmental activists, SHEPR offers only the most rudimentary acknowledgment of environmental concerns. Those entering the school without an environmental consciousness can "graduate" with their ignorance fully intact. Environmentalists and corporate interests come together to enact a project such as SHEPR. When working together by mutual consent, environmentalists' and corporations' value systems are not considered to be in conflict. But as the description above makes clear, these values are not shared evenly in the socialization project of environmental entrepreneurship.

What is at stake in in the School for Environmental Entrepreneurship is a *stratification of values*. SHEPR is not only a socialization project, it is also a value project (similar to a "racial project," in Omi and Winant 1986). It is a micro-interaction geared toward producing macro-level stratification between different social groups and their value-based claims. Such value projects target the collective assumptions that undergird those cultural norms that assign value. They render the stratification of values in a society as "common sense." The dominance of economic values requires a historical, iterative process of interpreting the nature-business nexus in favor of the latter.

Examples of the weighting of business concerns over environmental ones within SHEPR are numerous. Business experts who teach business master classes in the school generally have not integrated environmental concerns into the subjects they have been invited to address. Division of the two topics was the norm rather than the exception, in my experience. There were only three events that took place over the course of the five-day school that were explicitly environmental. Importantly, none of these activities awarded any points, which further underscores their lesser value. Student projects were rated according to a predetermined list of measures. At every reckoning, the environmental impact of a project comprised only one measure, while the entrepreneurial topics ranged from six to eleven different measures, each of which awarded points. A project that received no points for environmental stewardship could still outperform a project that received the highest markings for protecting the environment.

To understand the level of entrepreneurship's dominance over the environmental in SHEPR's design, it is helpful to imagine what the school might look like if the tables were turned. It is possible to conceive of a program that would take a potential business plan through an intensive incubator to strengthen its environmental record while giving a tip of the hat to its profit-generating potential. Rather than looking at the plan's budget, marketing, personnel, and logistics, the school could study its effect on water, air, carbon dioxide, toxics, and habitat destruction.

Perhaps the most important factor driving home the stratification of values between the environmental and entrepreneurial was the composition of the final jury. There was not a single environmentalist or environmentally knowledgeable individual on the final jury. As the program designers explicitly stated on the evening of the penultimate day, the final jury determines the outcome of the competition. The final jury collectively awards hundreds of points, more than enough to counteract the points earned during the school itself. The jury's composition, crafted solely of bankers and marketing executives, silently signaled to everyone present what is most important.

The result of this design structure dictated the outcome of the school, as well as the lessons students were likely to learn. Eco-Field was a raw materials extractive enterprise that received one of the lowest marks for environmental sustainability and whose co-leader continued to insist through the end of the program that Siberia's *taiga* was an inexhaustible resource. The project's expected profits alone (exemplified by the jury's physical reaction to them) solidified its status as the best business proposal in the school. Its poor environmental credentials paled in comparison to the projected 5 million rubles it would garner in its first year.

When I asked Elena Tvorogova about the jury composition, she replied that she wanted students to know what they would be facing when they actually look for investors. "If they want to start their own business, they have to make their case before investors and before banks," she told me, not environmentalists. This same pragmatism percolates up in her talks during SHEPR. As can be seen in her remarks on opening day, Tvorogova is always referencing the "real world" and what "real" entrepreneurs and investors are doing. Since environmentalists hold no sway in determining which businesses are funded in the "real world," neither would SHEPR grant them that privilege in their final jury. The unintentional effect of training individuals to navigate the "real world," however, is socialization into and perpetuation of extant social structures that may be less than beneficial even to environmentalists' own goals.

Variable Socialization: Ideological Degrees of Freedom

A pluralist society will necessarily be subject to value-based conflicts. The presence of conflict does not imply a pernicious form of stratification. Similarly, a socialization project does not necessarily equate to a value project. Socialization projects are always ideological. They always seek to promote some particular vision. But the scope of their claims can vary, as can the assumptions under which they operate, and these provide the means by which to evaluate their status as a value project.

Programs of socialization, education, and empowerment have *greater or lesser degrees of freedom*.[5] To better articulate how degrees of freedom operate in practice, let us consider two other environmental socialization projects being conducted in the Irkutsk region: an environmental summer camp designed and executed by En+ and "Eco-Schools," which is part of an international environmental education program hosted at Baikal Environmental Wave.

EN+ ENVIRONMENTAL SUMMER CAMPS

The En+ sustainable development department designed a summer camp for children living in its "mono-cities" [*monogoroda*], which are, essentially, "company towns" such as existed in the United States in the late nineteenth and early twentieth centuries. In acquiring old Soviet industries, En+ also became the chief employer in several Soviet-era mono-cities, and the company created an environmental summer camp for the youths living in them. The representative of sustainable development at En+ described it for me thus:

> This is not your usual kids' summer camp. It's a social game. They create an environmental city. . . . We hold elections, for mayor, or president, ministers, legislators: that is, all government, business, courts, society and so forth. So the goal is to form a proper model of a city where each person plays a particular role [*gde kazhdii zanimaetsya svoim delom*].
>
> Last year, our camp had an environmental theme. That is, businesses had to do an environmental project, like a waste treatment plant. First the kids come up with the idea, then they have to draw up the design of the plant. Next, they go before the legislature or the mayor and present it to these administrators [who have to approve the project] . . . The administrators should earn money from it—they receive taxes from each project. If they approve the project, the kids

have to make a real construction of it. The constructs are made of large wooden sticks. For this, they are trained in tying knots and in how to build a model . . . So we get a big exposition of these real constructions. Of course they are schematic models, but the kids explain what it is. They learn skills such as making a presentation.

All these projects have some skill involved that they will learn. For example, when it begins, the camp doesn't have anything. No discothèque or entertainment. Someone will become the PR director of the camp, and he will put on the discothèque. To do this, he needs to create a project—a business project—that includes the list of participants and explains their roles, bring it to the government for approval, and then they can begin to earn money. You get paid to conduct an activity and coordinate it. [He gets paid] with virtual money, not real money. It goes in his account. And each project, like if they want to go on an excursion to the shore of Baikal, they can't just go, someone has to be the initiator. He has to make a project on how to take the trip, where to go, how many people, for how long. All this is to describe it and defend it before the government officials who have to approve the project. Next he assembles a team. And then, when the project is completed, he receives a salary which he distributes to his employees, and so everyone accumulates points. There is a Minister of Sport who conducts exercises every day. There is a guard—police—who monitors safety in the camp and prevents fighting. In the event that there is suddenly some conflict between students, they go to court . . . So it's a direct model of contemporary society.

The sustainable development representative explained the camp to me by saying that, when students arrive, the camp has nothing. The students themselves must create the "environmental city" in which they will live. And yet, the terms of that city are already laid out for them by the camp organizers. They are told they must elect an executive, legislators, judges and bureaucrats. There is no opportunity to reimagine government, to experiment with different forms of rule, or even to divide the camp into different groups that select different governing structures for each group.

Similarly, they are told that whatever they wish to do—be it to host a discothèque or to lead a vacation tour to Baikal, they must create a profitable business that provides a service. Notably absent from En+'s "environmental city" are nonprofit organizations or public goods and services. Neither are students free to reimagine economic structures, to brainstorm how goods and services might be supplied to their "environmental city" other than through the capitalist marketplace.

Economic inequality is written into the camp's very structure through the virtual money students receive for their initiatives, given first to the "businessman," who then disperses the earnings to his "helpers" and to the administration. Curiously, "taxes" from these businesses go into the administrators' accounts. Although some sardonic observers may agree that this is, in fact, a direct model of the Russian state, it does not show its young participants the redistributive role of taxes in the economy. Finally, the system of "virtual money" that is allegedly merit based has very real rewards. As the En+ Representative explained to me:

> At the end of the camp, we want to give gifts to the children, but so as not to just give them stuff, we hold an auction. We teach them the skills of an auction: what it is and how it is properly done. So at the auction, those kids who earned some money—virtual, not real money—have the opportunity to win something. Someone will get a notebook computer, someone will get a pen.

The camp reaffirms and rewards the virtue of capital accumulation and, in so doing, reinforces a meritocratic outlook on economic inequality. Meanwhile, the most tenuous component of the "environmental city" summer camp that En+ created is its environmentalism.

ECO-SCHOOLS

Eco-Schools is an international program that grew out of the 1992 United Nations Conference on Environment and Development in Rio de Janeiro. Overseen by the Foundation for Environmental Education, the program operates through nonprofit organizations in thirty-three countries, including Russia. Baikal Environmental Wave is a host site for the Eco-Schools project, and Zinaida serves as the liaison and certifier for projects in the Irkutsk *oblast* [region]. There are seventy Eco-Schools in her jurisdiction, approximately half of the total number for Russia as a whole. She described the program for me as follows:

> [Once the school signs up for the program], they choose a topic. There are four topics: water, energy, climate change and waste; but if the school wants to choose some other topic, they can. Many choose "Healthy lifestyles," some choose "Biodiversity." . . . So, essentially, there are seven steps in this program. There is nothing specific spelled out, the program just stipulates that you need the first step to be the creation of an environmental council. That is, the whole school should be involved, but you need an environmental council with

representatives from students, teachers, parents, sponsors; basically, there just needs to be some kind of council. The council selects the topic on which the school will work: for example, water. Then, they study the situation: what happens to water in the institution, how does the school use water? So, kids learn where water is leaking, where it drips, where they use too much, these kinds of studies. Next, they draw up an action plan—what they can do to reduce excess water loss, that is, to reduce water use. Drawing up an action plan is not only necessary to know what has to be done to reduce water usage, but it is also the educational part. So the third and fourth steps, once the plan is already underway, is monitoring and evaluating the plan, that is, they constantly keep track of how it is working, evaluating where it is successful or unsuccessful, whether there is some component that should be added or taken out. Obviously, there should be some outcomes. Then, there is the integration of the selected topic in all educational courses. For example, you have to look at water in mathematics and physics and chemistry—what is it, and in biology and physical education and in some extracurricular activities. Water, water, water. That is, they need to study it a little bit. And then they need to share it with others. [They must publicize] that the school is dealing with this topic, what they learned, and with whom they cooperated. That is, they have to share with others about what they are doing. Maybe someone will be interested and something else will come out of it. And the last, seventh step is to prepare an environmental code for this topic. So they draw it, how to save water, or write about it, some have written poetry, some list concrete steps of the different ways [to save water]. They can do it however they like. And that's it!"

Participating schools are instructed to create an environmental council, but how it is structured, its means of decision making, and its role in the school-wide project is left up to the schools themselves. There is even flexibility on the choice of topic under the large umbrella of environmental concerns. The project attempts to empower youth by showing them their ability to tackle concrete problems and to enact measurable change—but how they choose to do so is self-directed. The terms of nature protection, or even for collective self-management, are not laid out a priori, but are in service of the overarching goal of, for instance, water-use reduction.

Both the summer camps and the Eco-Schools project are geared toward youth socialization and empowerment. Both claim to be teaching young people important life skills, and both profess an environmental focus.[6] Each may be considered an ideological project, but they differ dramatically in the *ideological degrees of freedom* they offer their young charges in crafting their respective projects.

Eco-Schools allowed variance in the *means* to further the environmental *end*. This contrasts with the En+ summer camp, which offered little coherence on the *end* of their environmental city but regimented the *means*—stratified class economies, capitalist business enterprises, and bureaucratic government. Environmentalism at the En+ summer camp is something that exists within the confines of its pre-established set of social structures, which are presented to students as immutable.

Eco-Schools has built into its design wide latitude in the choice of theme, oversight structure, and stated ends. The flexibility does not necessarily mean that the project will be better or that students will actually take on the opportunity provided to them to reimagine their structured environment. Teachers at various schools could dictate terms to the students, even if the program itself does not. But ideological flexibility is built into the program's structure.

Value Projects in the Field of Power

By funding value projects over other more benign socialization efforts, corporations like En+ can shape the field of power in their own favor. When used in this way, corporate sponsorship is a power play. It works to curtail the generalizable power of another player in the field. The power play operates by using value projects to promote the ascendency of business values over the environment and limit legitimate environmentalist claims in the public at large. The value projects restrict the scope of "worthy" activity and thus limit the scope of civil power. As long as business values are ascendant, civil society cannot threaten business as a whole without losing civil power.

The financial power of corporations allows them to make such a play because sponsorship grants them access to the activities of the nonprofits that they sponsor. Not all sponsorship is a power play (see Chapter 5). But sponsorship can become a power play when it allows corporations to selectively fund only those socialization projects that double as value projects, as did SHEPR. We can see the power play when we look at those socialization projects that En+ did *not* fund.

In an interview I conducted with the Tvorogova after the conclusion of SHEPR, we addressed the topic of "unfunded" projects.

KATE: So you already had the idea [for SHEPR] before En+ [came], and you said, "Here, I have an idea that I would like [to do]."

TVOROGOVA: Actually, we suggested to them a choice of several ideas. They chose this one.

KATE: That means that you had other projects that you wanted to do, but did not get the money for.

TVOROGOVA: Yes.

KATE: For example? Shall we dream?

TVOROGOVA: Let's dream. There is a long-standing dream that is shared by everyone in the [organization] . . . [to create] a Center for Alternative Technology. That is when, on an actual plot of land, there stands an actual house, which uses functioning alternative technologies . . . So this center—on the one hand, it's a testing ground, on the other hand it's an educational center, where people can come and hear all about it, see it, hold it, touch it. This is a very important point. They can see the calculations: so if we cultivate the land in this way, then what we get here is such-and-such results. And it costs this much. Same with the building itself: [this is what happens] if we stand solar panels and use energy-efficient materials there and such. [In Russia] such experience is catastrophically lacking. That is, we already have a business structure on the market that could be promoting this economy. But we can't just coerce businesses to do it, so it is necessary to create these centers. Because to promote these technologies and these ideas to consumers—that is, to actual purchasers—people need to see for themselves with their own eyes that it works, and that it is sufficiently effective. [The center is necessary] so people can come once, have a look, leave, think about it more, then come back again. Do you understand? It's a complex process to change consciousness, to change priorities, etc. So my idea was precisely such a creation— one of the projects that I proposed—was the creation of a Centre for Alternative Technology somewhere in the southern Baikal area.

Like her fourth-wave counterparts in the West, Tvorogova's environmentalism seeks to harness the power of the market to shift practices of production and consumption into something sustainable. Her greatest dream is a socialization project that would allow consumers to gain hands-on experience with alternative energy and energy efficiency. With a wide array of products and real-time comparisons in outcomes, her vision would have ideological degrees of freedom while drawing attention to environmental protection.

En+, on the other hand, was not ready to support this type of values project. The company is glad to make minor accommodations for environmentalism, but to reimagine power sources or to limit the use of natural resources is not a positive goal for them, a fact that was made clear to me in my interview with the representative of sustainable development at En+ in Moscow. Throughout our conversation, she was bright, pleasant, and enthusiastic as she described the multitude of programs sponsored by En+, both in Irkutsk and elsewhere. There was only one moment when this sunny façade was broken.

> KATE: Of course, there is a huge international conversation about global warming. I'm curious how you would answer environmentalists who said that your company is selling coal to China and . . .

At this question she visibly bristled and shot back:

> EN+ REPRESENTATIVE: Where do you live? In a building? Do you use light? Do you use dishes? Cups? Spoons? Do you want to live in a tent? We produce light. We make, what? Coat-hangers! Everything that is metal. It seems to me that the real question is that large industrial enterprises are the main source of environmental harm; that's alright, that's understandable. In order to do this business correctly, we're going to engage environmentalists from the point of view of industry. How can we properly industrialize, so that plants can work better? And we are trying to build these processes at our factories from a more correct environmental point of view. For that reason, we call in [environmentalists]. . . To an environmentalist who says—"Close the plant!" I say, "Fine, go live in a tent. Eat with your hands. You're not doing that." . . . So, I believe that you must be reasonable.

The En+ representative essentially breaks down environmentalism into two camps: the reasonable and the unreasonable. In doing so, she defines the boundaries of legitimate environmentalism. Environmentalism is good when it helps En+ make its practices more efficient, but environmentalists are not to call into question what those practices should or could be. They can "green" what is already being done, but they have no legitimate voice if they question whether contemporary practices *should* be done. Promoting alternative pathways toward economic development is beyond their legitimate sphere of influence, according to the representative from En+. Environmentalists can assist but cannot hinder the course of action that the company chooses for its own purposes.

Yet, because En+ is footing the bill for environmentalist work, it is in a position to render its own definition of legitimate environmentalism dominant. In activism as elsewhere, the question is not simply whether resources matter in movement mobilization (McCarthy and Zald 1977). Less acknowledged but equally important is the question: *whence* those resources? The source of one's money has an impact on the ideological content of one's activities (see Oreskes and Conway 2011; Berman 2014, Farrell 2016). For example, GBT activists were concerned about their autonomy, despite the fact that En+ only paid for projects that GBT would have conducted anyway. In purchasing the glamour of civil power, autonomy is necessary and assured. But socialization projects are a different beast, and the ability to selectively fund desired projects opens the door for sponsors to shape the legitimate terrain of social action. It is precisely here—in projects of socialization—that the question of autonomy truly matters.

Conclusion

Value projects that delimit relative "worthiness" of social goals also limit the reach of civil society's power. Public support for civil society groups wanes as public approval for their activities and demands weakens. The trick behind the value project is to maintain the worthiness of a social organization's cause, while simultaneously ensuring that its value is generally considered lower than the values of their opponent in the field of power. Economic elites, such as Deripaska and his partners in En+, aim to foster a stratification of values in the public at large between the requirements of business and the requirements of nature, such that nature is only protected insofar as it conforms to business needs. When business logic becomes the socially constructed limit for environmental concern, the power of civil actors is similarly constrained by that same business logic. To do otherwise would run counter to the public assessment of virtuous activity, and it would weaken any civil society group that steps too far out of bounds.

Corporate sponsorship is the means by which this power play takes place. Civil society actors frequently foster socialization programs to help achieve their aims. Economic elites that give money to support social organizations may choose certain types of socialization programs over others. Namely, they sponsor those programs that bolster the corporate position in the field of power and that mitigate the threat of civil power by constraining its range of socially worthy demands. The result is a kind of shield or ideological buffer that protects business. Civil society may be strong and independent, but it becomes weaker precisely when it seeks to deploy its power against standard business practice.

Not all corporate sponsorship involves value projects.[7] Value projects are only those socialization projects that attempt to stratify social values. Socialization projects themselves come in many forms and can be enacted by myriad social groups. All socialization projects are vessels of ideology, but they offer different ideological degrees of freedom. The extent to which a socialization program is also a value project can be assessed, in part, according to these degrees of freedom. How many assumptions must be made in order to participate in a project? How much attending structure is required beyond the stated aims of the project? How much flexibility is provided for self-determination within the project goals? Attending to these differences opens the door to a greater understanding of value projects as a specific form within the more general practice of socialization.

The field of power analysis I have provided from Russia also sheds new light on fourth-wave environmentalism in the West. The continual enactment of value projects stratifies social values in favor of business. Once so stratified, environmentalists cannot easily succeed in their aims without cooperating with corporations because they lose their power when they directly oppose them—or at least their logic. As ecological modernization theory continues to advance, scholars may use their appreciation for the field of power and their understanding of value projects to provide a more nuanced critique of the relationships between environmentalists and corporations into the future.

En+ is not the only actor in the field of power that seeks, or has sought, to utilize social projects in this manner. The Communist Party in the Soviet Union similarly sought to frame all social activity such that it corresponded with party ideology. And the state under Putin has revived the tactic, investing in youth projects that, while polyvocal, nevertheless seek to promote an apolitical volunteerism that does not threaten state power (Hemment 2015). But the state has a far greater arsenal at its disposal when it comes to making plays in the field of power. Socialization projects are petty endeavors compared to the strength of the law. While the law is not the only form of generalizable power, it has an oversized influence compared to its peers. When the state decides to enter the game, the consequences in the field of power are much less subtle.

7

State Suppression of Baikal Activism

Every year, Irkutsk is host to the "People and Environment [*Chelovek i Priroda*]" film festival, an international festival that specializes in documentary and popular science films involving the natural world. At the close of one of the roundtable sessions at the festival in 2012, the German director of a featured documentary asked to say a few words. He addressed the audience in English, and Lusiya, a volunteer with GBT, translated his comments into Russian.

> If I may, from my experience, offer three pieces of advice . . . First, decrease demand for energy. This is something we all can do, it is a way we can all participate, especially the poor. Decreasing demand also saves money on energy, and everyone can do it. Second, focus on coal. This can be done where you have district heating, such as you have here . . . Finally, if you are going to go the political route, then you need to have a feed-in tariff law. You must elect representatives who support this law. Those countries where you see change are those where the people voted *for* people who work for change and voted *out* those who did not!

His comment was met with chuckles, smirks, and exchanged glances in the audience. Lusiya, while translating this well-intentioned advice into Russian, looked embarrassed. She added her own addendum to the translation: "I know that there are some people smiling out there, and we all know why. But in Germany they have a very strong Green Party, and they were able to vote for them and create these laws."

The director then closed with an emphatic line: "If you believe in democracy, you have to do this."

With a pained expression, Lusiya translated his closing statement. People in the audience started to laugh. One young man laughed ostentatiously, slapping his knee, and then got up and left the auditorium. Lusiya turned to the director and said in English, "In case you are wondering what is going on here, you have to understand that democracy is a painful question in Russia."

The Russian State in the Field of Power

Democracy has never successfully taken root in Russian soil. For some observers, this is due to an intrinsic aspect of Russian national culture that demands a "strong leader" rather than wide participation (Keenan 1986). For others, authoritarianism is like a bad habit that Russian society keeps returning to for want of alternate experience (Figes 1996). It is a well-trodden path to which it is easy to find one's way (Paxson 2006).

The field of power provides new insights into this question of "Russian exceptionalism." From the field of power perspective, democratic governance under the rule of law is an accomplishment rather than a right. Far from being normative principles to which state actors naturally and willingly adhere, these state structures and legal restrictions show the success of previous power plays by opponents in the field. Once put into place, democracy and the rule of law help to ensure that state power does not become overbearing and that the field remains relatively open to alternate power players. But democracy and the rule of law are still accomplishments, and their successful implementation depends very much upon the capacity of alternative powers at key historical junctures. The tight fist that has held state power in Russia since the days of Empire has left little opportunity for alternative powers to expand the field. Social and economic crisis has accompanied every major transfer of state power in Russia, and the state has successfully recouped its former dominance over the field of power. The present government under Vladimir Putin is no exception.

Of all the generalizable powers in the field, the state plays a unique role. As the sole sovereign over a particular territory, all social actors under its dominion are accountable to its dictates. The state sets the rules that govern all social fields within its purview, including those of business and civil society. When considering the field of power and the arsenals available to actors within the field of power, control over the apparatus of the state may be the ultimate trump. Economic elites seek to use their power source—money—to alter the legal system by financing political campaigns or buying off corrupt state officials. Actors in civil society likewise seek to control the legal apparatus, by using civil power to mobilize movements and encourage the public to withhold its consent. But these tactics are, at best, a mere proxy to wielding power over the rule-making machine. Within its sovereign territory, the state has primacy of place.

But state power is limited by geography, and the global dimension of the field of power adds new complications. The era of globalization is not a neutral phenomenon for authoritarian governments; to the contrary, the force is decidedly threatening. While civic and economic power may be strengthened by globalization with its access to wider markets, social networks, supplemental resources, and imagined possibility, the state has a more ambiguous relation to the global field of power. External political players can and do seek to influence or pressure other sovereign states through incentives or sanctions, but a state's territorial sovereignty generally remains inviolate. With the rise of globalization, however, there is greater need to legitimate state authority in the global community, and citizens are more likely to judge their leaders based upon global benchmarks. The resulting "isomorphism" in governance may be a boon or a burden for domestic subjects, as these isomorphic laws can be seen in such a wide array of areas as human rights laws (Tsutsui and Wotipka 2004) and austerity measures (Babb 2003).

In the immediate years following the Soviet collapse, there was tremendous faith in these isomorphic pressures. Scholars spoke with a sense of teleology when discussing Russia's trajectory: the country was in "transition" away from one-party rule and a command economy, and toward the Western liberal-democratic ideal type (e.g., Sachs and Woo 1994). During the 1990s, when the state had lost its legitimacy and the economy had been shattered, the Russian government was responsive to international pressure. The allure of participation in global governance institutions like the World Trade Organization and the G8 helped steer a common course for Russia and the West. However, these relationships did not cure the "permanent crisis" (Shevchenko 2008) that characterized Russia in the 1990s, and when Putin returned the state to solvency and stability, it was done in part through a repudiation of those isomorphic pressures the West had been exerting.

Since then, Putin's Russia has walked the line between global participant and resistant renegade, between team player and maverick. While Putin can usually find a globally normative frame to justify and legitimate his choice of action,[1] he clearly understands that his particular power source is tied to state sovereignty—and that globalization is uniquely disadvantageous to his own influence over the field of power.

Environmentalists at Lake Baikal have long been a thorn in the side of the state. Neither have they been immune to state repression. But in 2012, the Russian government made a power play specifically targeting the power of Russian civil society. The Foreign Agent law was designed and implemented in such a manner that threats to the state could be targeted and summarily disposed of. Its two-pronged attack both limits civil society's scope of activity and simultaneously seeks to sever its global connections. The law creates a fundamentally new shape for the Russian field of power and provides a means for maintaining state dominance in the modern, globalized era.

Threats from Below

On December 4, 2011, the Russian Federation held its quinquennial legislative election amid simmering, albeit subterranean, political discontent. Putin's popularity was falling, although still well above the 50 percent mark. Frustration with corruption and bureaucracy was percolating through a middle class now grown accustomed to Western business practices. At the United Russia party convention, one-term president Dmitri Medvedev announced that he would not seek re-election so that Putin could take on the title for a third term. The backdoor decision to have no primary election in the overwhelmingly dominant ruling party suggested that the presidential election was already over before the campaign, and Putin's re-ascendancy was a fait accompli. The move angered many Russians, whatever their personal feelings about Putin, because they felt they had been denied the right to select their own president. These feelings of dissatisfaction and betrayal were brought to the ballot box for the parliamentary elections in December 2011.

In a fair contest, there is little doubt that Putin's party, United Russia, would still have taken the majority of seats in the legislature. Nonetheless, the election was highly suspect. There was ample evidence of elections fraud and over a thousand official reports were filed citing voting irregularities (Schwirtz and Herszenhorn 2011).

Opposition groups began protesting on election day and continued throughout the week, culminating in a massive demonstration on December 10, 2011. Tens of thousands poured onto the streets of Moscow for what was the largest protest since the collapse of the Soviet Union. The demonstrators were a diverse group, coming from different age levels, social classes, and political affiliations, but they were united in their call for fair and transparent democratic elections and the end of Putin's long reign (Barry 2011).

Elites in United Russia, including Putin and Medvedev, made crude and disparaging remarks about the protesters and their leadership in an attempt to ridicule them and undermine their legitimacy (Elder 2011). But these comments only seemed to further enrage the movement. United Russia began calling on its supporters to pose countermovement mobilizations (Schwirtz 2011). Anti-Putin demonstrations continued through the winter, with protesters braving subzero temperatures to publicly display their discontent. The state did make some (short-lived) concessions to the protesters, but demobilization in March 2013 was not the result of victory. Putin handily won re-election for a third six-year term, and demoralized protesters simply stopped showing up to scheduled rallies. As opposition numbers diminished, the monthly demonstrations became increasingly marred by violent confrontations between protesters and police (Elder 2012). Leaders' homes were raided (Lally 2012). Mass arrests, including of opposition leaders, also worked against the continued mobilization. New laws curtailed the freedom of assembly (Bryanski 2012). By midsummer, the "Russian Spring" was over with little to show for its efforts.

Threats from Abroad

For some, the winter protests in Russia were reminiscent of the "color revolutions" that swept through the former Soviet bloc in the early 2000s, a comparison particularly pointed when demonstrators adopted the color white as their unifying symbol. In 2002, pro-democracy protesters overturned the established ruling regimes in the "Orange" (Ukraine), "Rose" (Georgia), and "Tulip" (Kyrgyzstan) revolutions. Many remaining post-Soviet governments, including Vladimir Putin's, were concerned that their regimes could meet a similar fate.

Importantly, ruling elites in the former Soviet Union did not think that the color revolutions were simply the expression of domestic discontent, but rather that their citizens were being manipulated and motivated into action by foreign elements. Namely, they professed that Western governments were making use of NGOs to enact their values and preferences domestically (e.g., Traynor 2004). Nonprofits became suspect in the minds of elites as geopolitical stool pigeons. Among the main culprits called out were George Soros's Open Society Institute and USAID, both well known for funding pro-democracy and human rights organizations in Eastern Europe and Central Asia.

Partly in response to the color revolutions, the Putin government passed a law in 2006 that sought to restrict foreign influence on the NGO sector. In its original form, the law would have created a separate status for registering and monitoring NGOs that received foreign funds, but in the face of domestic and international pressure, the law was softened. Still, it imposed tough audits and reporting requirements on domestic civil society organizations and gave the state the right to deny official registration to organizations whose mission appeared threatening or whose personnel the state deemed problematic. It also allowed the federal government to tax grants from foreign entities at a rate of 26 percent, which severely limited the number of foundations willing to operate in Russia[2] (Crotty, Hall, and Ljubownikow 2014). Several of the more restrictive aspects of this law were repealed in 2009 under the Medvedev presidency, but registration and auditing have remained a complicated bureaucratic difficulty for many Russian NGOs in the early twenty-first century.

The 2012 Foreign Agent Law

One of the more lasting effects of the pro-democracy/anti-Putin demonstrations that began in December 2011, and the embodiment of the Putin regime's apparent fear of threats from abroad, was the so-called Foreign Agent law. The law was adopted by the Duma on July 13, 2012, and approved by the Federation Soviet on July 18, 2012. President Putin signed it on July 20, 2012, and it went into effect 120 days later on November 17, 2012.

The Foreign Agent law is not simply a new means of regulating civil society: it is a power play by the state against civil society. To understand the law as power play, it is helpful to remember that civil power comes from its capacity to mobilize the public through its recognized worthiness. We must also remember the important role that transnational activism plays in helping the general public to imagine alternatives and believe in their ability to work toward them. The Foreign Agent law takes on each of these sources for civil society's strength: capacity, worthiness, and the imaginative possibility that grows from global connections.

The text of the law requires nonprofit organizations "receiving cash and other assets from foreign sources . . . and participating in political activities carried out in the territory of the Russian Federation" to register with the state as a "foreign agent."[3] Doing so gives the state far more reach and oversight into the operations of that organization. Nonprofits in Russia already have strenuous reporting and auditing requirements, but "foreign agents" are subject to additional scrutiny. They must report to the state "the amount of cash and other property received from foreign sources . . . , the purposes of expenditure of these [assets] . . . , and their actual spending and use." (Federalnii zakon ot 20.07.2012 g. N 121-F3).

These reporting requirements not only give the state a great deal of insight into the operations of particular NGOs, they also add a significant burden in time and organizational resources. As the law states:

> Non-profit organizations that act as a foreign agent shall submit to the competent authority: documents containing a report on their activities and on the personnel composition of its governing body every six months; documents on the goals of expenditure for the use of cash and other assets, including those received from foreign

> sources—quarterly; and, finally an annual statutory audit (Federalnii
> zakon ot 20.07.2012 g. N 121-F3).

These various reporting schemes—quarterly, semiannually, and annually—greatly exceed the already strenuous reporting requirements for regular NGOs, with money, time, and resources diverted from projects toward the bureaucratic maintenance of the organization itself, which can itself dampen mobilization and protest (Piven and Cloward 1977). They plague organizational capacity.

More troublesome than the bureaucratic reporting of income and expenditure is the increased governmental oversight of organizational activities outlined in the new law. The law states that, for registered foreign agents "to participate in political activities in the territory of the Russian Federation, [they] shall, prior to attending said political activities, apply to the [regulatory authority]" (Federalnii zakon ot 20.07.2012 g. N 121-F3). In other words, the state has essential veto power over the activities and actions of NGOs, particularly those that may be deemed "political," broadly defined. The law gives the state an ability to suppress assembly and speech that is unprecedented since the collapse of the Soviet Union. A foreign agent cannot legally act without the state's approval.

Equally troubling, the law is deliberately designed to portray civil society as the Other, designating certain segments of civil society "un-Russian." Not only are these domestic groups classified as "*foreign* agents," but they are required by law to have the moniker attached to all their public materials. As the law states:

> Materials published by a non-profit organization acting as a foreign
> agent, and/or distributed by it, including through the media, and/or
> by use of the information and telecommunications network called
> "the Internet," must be accompanied by an indication of the fact
> that these materials are published and/or distributed by a non-profit
> organization that fulfills the functions of a foreign agent. (Federalnii
> zakon ot 20.07.2012 g. N 121-F3)

Organizational representatives of designated nonprofits cannot speak publicly without making it known that they are an agent of a foreign interest. Russian citizens' groups can have their voice de-legitimated in the domestic public sphere by the requirement to always attach the moniker "foreign" to their "free speech." The requirement is an attack on the "worthiness" and virtue of the targeted organization, the effect of which cripples such a group's ability to wield civil power.

The law is also sufficiently punitive. It establishes certain conditions that entitle the state to conduct unscheduled checks and audits of NGOs who are suspected of acting as foreign agents. If a nonprofit is found to be in violation of the law it can have its operations suspended for up to six months. Organizations and their leadership who do not comply with the law governing the registering and monitoring of "foreign agents" are subject to a fine of up to 300,000 rubles, up to four hundred hours of compulsory work, or imprisonment for up to two years. Should the restrictions on capacity and the attacks on worthiness themselves fail, the state can simply and summarily eliminate its opponent from the field.

The law is directed toward maintaining the state's dominance in the domestic field of power, since its sovereignty stops at Russia's borders. The moniker "foreign agent" is designed to diminish domestic support for labeled NGOs. But bundled into the mix is a very real attempt to prevent Russian civil society from utilizing the global plane as a power player. It specifically targets foreign funding, but the aim is far broader. Along with the giving and receiving of resources come interaction; transnational affinity; international solidarity; and exposure to new ideas, alternative views, imagined possibility, and even— potentially—belief in change.

The Turn of the Screw

Despite this potentially worrisome and problematic new legal framework, the nonprofit organizations I studied in Irkutsk were generally unconcerned about the new law when it went into effect in the fall of 2012. To the extent that they thought about it at all, it was in relation to the anti-government protests six months prior. They generally assumed that the law had been created to target Golos, the election-monitoring organization whose work helped spark the December 2011 uprising with evidence of widespread election fraud. Although I knew that both GBT and the Wave were recipients of foreign funding, neither considered the law to apply to them.

The reason was a paragraph in Article 2, Section 2 of the law that listed certain exemptions and exclusions. Some organizations were simply not covered by the law, and these included religious organizations, political parties, government-affiliated organizations (e.g., VOOP), corporate-affiliated organizations (e.g., Volnoe Delo, Defend Baikal Together), chambers of commerce, and international nonprofits that are registered abroad (e.g., World Wildlife Fund, Greenpeace). Other exclusions are found in the definition of "foreign funds" and "political activity." While the definition of foreign funding is fairly self-explanatory, the definition of "political activity" is both broad and vague. The law defines political activity as activities conducted with a goal "to influence the decisions of public authorities, aimed at changing their work on public policy, as well as to influence public opinion for the same purpose" (Federalnii zakon ot 20.07.2012 g. N 121-F3). Essentially, organizations who receive any money from abroad cannot also attempt to influence public opinion or policy without being designated a "foreign agent." However, the law also states that certain activities are excluded from the definition of "political activity," and one of these is "the protection of flora and fauna."[4] The organizations I studied in Irkutsk, as environmental organizations immediately concerned with the protection of flora and fauna, saw themselves as exempt from its oversight. For one of them, this belief proved to be fatally incorrect.

MARCH 26, 2013

"I heard on the radio this morning that they are cracking down on NGOs," Jennie said to me from behind her computer while we were working at the office of Baikal Environmental Wave. "They had a representative from three organizations on a talk show. The prosecutor is accusing them of being foreign agents. There are some others, too. Only one is an environmental organization—one I've never heard of in Krasnoyarsk."

"You think the Wave will be next?" I asked her.

"Maybe!" she answered laughing. "We'll see," but her tone suggested that her concern was more intellectual than existential.

APRIL 1, 2013

In the afternoon, the Wave received a phone call from the prosecutor's office, alerting the staff to a coming fax. The fax, which arrived at 3:44 p.m., stated that, in fulfillment of the law "on combating extremism" in public, religious, and other noncommercial organizations, the nonprofit was being "checked." The letter demanded that the Wave provide the prosecutor's office with copies of 1) their organizational charter and 2) "documents confirming the sources of funds and other assets (from which organizations, from which actual persons), their nationality, the general purposes for which these funds were allocated, the total amount of money raised by the organization in 2010, 2011, and 2012, and on which goals the funds were spent." The letter stated that the documents were to be delivered to the prosecutor's office by April 2, 2013. The date was printed in bold and underlined. "By 11 a.m." was scrawled in pen beside the date.

With less than twenty hours to collect the requested paperwork (and most of those after business hours), they set to work, prepared to burn the midnight oil. The Wave staff was up until 1 a.m. photocopying documents, but managed to compile the needed materials in time for the 11 a.m. delivery.

"Success," the Wave announced on its website, following the midnight photocopying frenzy. The posting even took the occasion to sympathize with the staff at the prosecutor's office, reminding readers that "they, too, are now hard at work and facing sleepless nights."

APRIL 4, 2013

In the office, everyone was back at work, as though the *proverka* [check] had never happened. The Wave's staff was already busy focusing on its latest projects. Marina was planning the conference that would be the culmination of the Tahoe-Baikalsk-Goloustnoye grant. Artur and Zinaida were hauling pieces of the Interactive Education program to the Siberian Convention Center for an education fair to which they had been invited.

"Are you worried about the *proverka*?" I asked Jennie as we sat behind our respective computers.

"No," she replied. "I don't think they will call us foreign agents."

"Why not?" I asked.

"Because they know that if they do, it will make a big stink!" she said smiling.

APRIL 11, 2013

At noon, the prosecutor's office called after sending a fax to the Wave office, requesting additional materials. Specifically, the Wave was to collect and copy:

1. Grant agreements, invoices, payment orders, and reports provided to foreign funders
2. Statements of the goals and objectives of workshops funded by foreign sources (handouts, presentation material, and presentations made at the workshop)
3. Clarification on whether Baikal Environmental Wave conducted public events (actions, protests)
4. Information on whether Baikal Environmental Wave prepared and published printed materials, media articles (including on the Internet), and other media projects (indicate which documents were produced by which activities), and copies of the published material
5. Copies of orders for business trips make by the Wave staff
6. Accounting documents confirming the costs incurred on business trips, and all travel reports
7. Reports to foreign funders on the progress of project implementation

Baikal Environmental Wave had to assemble the above list of documents for the years 2011, 2012, and 2013 and deliver them to the prosecutor's office. They were given five hours to comply.

When I arrived at the Wave office there were three heads bent down over three desks, keyboards tapping, papers shuffling, and an unusual and tense silence. For the first time in all my visits to the Wave office, the overhead lights were on during the day.

"I brought some sugary energy," I told Marina, offering her a package of merengue cookies.

"In a moment," Marina smiled, and I set them on her desk.

Upstairs, Katya, Artur, and Jennie were sitting around the table, having just finished lunch.

"I think Putin is crazy. I can't explain it otherwise," Jennie said to me in English. "How does he think this will go over in this modern world? Is this really how a serious country acts, spending its time, resources and energy on small nonprofits like the Wave? How can anyone take this seriously, take Russia seriously? I heard on Ekho Moskvy someone suggesting that NGOs simply register as foreign agents and be done with it. But I don't think that is the right track to take."

"Why not?" I asked.

"Because it is not true!" she answered emphatically. "It's a lie! I don't want the Wave to sign something that is a lie! And besides, what happens the next time they want to put an oil pipeline by Lake Baikal, and when we say no, they will say we are foreign agents, sabotaging Russia's advancement!"

At this point, Zinaida came up and asked for help finding a receipt for one of her projects. We all headed downstairs.

"Marina, you still haven't eaten lunch," Jennie said. "Go eat something."

"I'll eat at home," Marina answered placidly.

Now there were seven heads bent over seven desks. Sounds resumed: the whir of the printer and the chink-chink of the Wave's official stamp on duplicate documents.

"I can't find the report on the bio-toilets," Artur called out. Meanwhile, there was a trading of places while Zinaida moved to Artur's computer in order to print. Temporarily displaced, Artur paced the room. It was a tense game of musical chairs.

Suddenly, the office was filled with the electronic tones of a Beethoven sonata—Vera's phone was ringing.

"Hello? No we are still working, gathering everything," she said to what was evidently the prosecutor's office on the other end of the line. "There is no way we will finish by 5 p.m. That's in fifteen minutes. How late will you be in the office?"

The prosecutor's agent apparently suggested the Wave just bring them the originals, but Vera would not allow it, saying the originals were not to leave the office. Artur said he would be willing to take the originals and stay all night watching the prosecutor's staff photocopy them. At a quarter after five, Vera and Artur left with what they had managed to collect thus far, planning to finish the rest the following morning.

APRIL 12, 2013

Jennie was still smarting from the *proverka*. She continuously bemoaned the waste: the wasted paper, ink cartridges, and time. It was evident that she was fighting her natural instinct toward civil disobedience.

"I wanted to fight," she said, "but I had to step back. I can't put my neck out there because, at the end of the day, I'm not the one who would get hurt. Vera, Artur, and Lev are the co-directors, so if it comes down to it, they would be the ones to go to jail, not me. They would be the ones with a huge fine. Of course, the organization will support them with that, but still."

When she learned later that certain other organizations in Russia in her social network had, in fact, decided not to comply with the *proverka*, she worried that she would be judged harshly by her NGO peers for capitulating.

"[But] there is a big difference between submitting documents and registering as a foreign agent," she said, comforting her conscience. "We'll never do the latter."

Vera, meanwhile, had been helping a woman from the prosecutor's office create a table of the Wave's various grants. Apparently, the woman informed Vera that the Federal Security Service (FSB) was unhappy that the Wave had mentioned the *proverka* on social media.

APRIL 15, 2013

A three-page letter arrived by post to the office of Baikal Environmental Wave. It was from the office of the prosecutor. Its conclusion: the organization is a foreign agent and must register as such. The Wave was given one month to respond.

APRIL 16, 2013

I learned about the letter the next day from Jennie Sutton.

"I thought you said they wouldn't do such a thing because of the big stink it would cause," I said.

She paused for a moment. "I was mistaken."

APRIL 18, 2013

When I arrived at the Wave office today, I noticed something unfamiliar about the building front and did a double-take. Scrawled across the building in black spray paint were the words: "Foreign Agent ♥ USA" (fig. 7.1).[5]

I hurried inside where I was greeted by Jennie and Yulia.

"Did you see our new decoration?" Jennie asked, with a laugh.

"I did!" I said "It's incredible!"

"You saw it?" she confirmed, "Because we all walked right by it. Zinaida was the one who noticed it."

"I took a picture of it," I said.

"So did we!" she said. "It's already up on the Internet. We called the police, and they were here, writing everything down. And the TV station has been out. And another is coming. Of course we will paint over it, but we have to wait until it is all reported."

I handed her the cookies I had brought as a gift, and Artur said, "Uh-oh. That's help from a foreign agent," with a mischievous grin.

I learned later that the exact same phrase had been spray-painted on the front of Memorial, a nonprofit organization in Moscow (fig. 7.2).

Figure 7.1 The office of Baikal Environmental Wave in Irkutsk with "Foreign Agent ♥ USA" graffiti. Photo Credit: Author

Figure 7.2 The office of Memorial in Moscow with "Foreign Agent ♥ USA" graffiti.
Photo Credit: Yulia Orlova/Memorial

APRIL 25, 2013

When I came into the office, I heard that Reviving Siberian Land is now being checked by the prosecutor. While everyone sat at the table for tea, Elena Tvorogova came into the room looking pale and exhausted. She was invited to sit, but she refused, saying she had been sitting all day. Instead she stood, rocking back and forth, foot to foot, while she nibbled on sliced fruit. Once we were alone, I asked her about the *proverka*.

She said she had finally finished delivering a stack of papers "this big," holding her hands about two feet apart and making a frustrated face. I asked whether she had counted the cost of all that paper and ink.

"It's not the paper and the ink," she dismissed with a wave of the hand. "It's the time! All those things that I couldn't do while I was having to do that."

"What do you make of it all?" I asked her.

She paused, then said, "It isn't pleasant. I love my country, my region. I love Baikal. And to have the prosecutors spending their time and resources looking at me . . . There are *real criminals* out there, people doing really bad things, but the prosecutors are looking at me."

She was staring out the window while she spoke with a sad expression, as though it sat heavily on her shoulders that the country she loved and worked for would repay her efforts thus.

The text of the Foreign Agent law was clearly intended to curtail the power of resurgent civil society. But text was not the only means by which law was used against the Wave. The manner in which the law was implemented and enforced also signaled the power of the state and the relatively powerless position of civil society vis-à-vis the law. The *proverka* seemed designed to repress and intimidate rather than merely implement a federal law. The Wave was asked to compile three years' worth of documents overnight, and then again in five hours' time. There was no legal justification for requesting such a quantity of materials with such a short turn-around; the timeframe alone guaranteed that the organization would fail to fully comply. The prosecutors' reference to the FSB is suggestive, as is the comment's unveiled attempt to limit the Wave's public voice on the foreign agent law.

Perhaps the most disturbing of these intimidation tactics was the slogan sprayed across the office storefront. While no hard evidence can be mustered to tie this act of vandalism to the state, the Wave was convinced that it was not the work of a rogue hooligan. First of all, the Wave had only been notified of the prosecutor's decision two days before, and they had not reported it online or in the media. Public knowledge that the Wave had been determined a "foreign agent" was limited. More telling still is the fact that the exact same phrase had been spray-painted on the office front of another targeted NGO thousands of kilometers away from Irkutsk. If indeed conducted under the auspices of the state, such a tactic would suggest extra-legal intimidation and harassment, designed to persecute or frighten the Wave members into conformity with the state's agenda.

Theater of the Absurd

While Irkutsk NGOs struggled to understand and comply with the new law and the accompanying *proverka*, there were a number of questions that were circulating among activists that could only be answered by authorities. For example, why were NGOs required to provide documentation of foreign funding that occurred prior to the passage of the new law in 2012? Why were environmental organizations being checked at all if activities "protecting flora and fauna" were considered exempt? I had questions of my own regarding the order that organizations were checked, and why some were checked and others were not. Among the NGOs and their online discussion boards and listserves, stories fluttered around, full of rumor, speculation, and supposition. If answers were to be found, they would not come from the NGOs but from the prosecutor.

MAY 6, 2013

I walked to the city prosecutor's office, arriving just as it opened at 9 a.m. The office is located in the center of town on a pedestrian street surrounded by shopping boutiques. After opening the heavy outside door, one is immediately greeted by dark, heavy stairs with a thick wooden railing. At the top of the stairs, a heavyset young woman in uniform was perched in a guard booth. I explained to her who I was and what business I had there. Her eyes did a fluttering roll when I mentioned the *proverka*, but she took my passport information for her record. The windowless lobby for the city prosecutor was small, dark, and drab. Old brown sofas lined the beige walls. More than half the bulbs were burned out in the overhead chandelier and the wall lights. After logging me into the directory, the security guard directed me down a dim corridor and to the second door on the right.

Entering, I met three women in their thirties, chattering happily as they removed their coats. The room was somewhat cheerier than the lobby, with two long windows illuminating cheap desks and tables covered with boxes and stacks of papers. The women looked up at me with surprise. I repeated for them who I was and why I had come. At the word *"proverka,"* they also rolled their eyes. They told me to return in an hour.

When I returned, I was taken to a different room that was occupied by two other women who sat at desks that were facing one another. I told them that I was a researcher, studying environmental movements around Baikal, that a few of the organizations in my study were now being checked as foreign agents, that there were many questions that local activists were asking, and that I hoped to hear how the prosecutor's office would answer them.

The meeting lasted approximately twenty-five minutes and was only productive as a lesson in evasion. Much of the time, the three of us were all talking at once. I would start a question, but before I had finished, the two women would interrupt and speak back at me. When I asked a question, they would give an answer that did not fit the question. When I would repeat the question, they would say that they could not answer it. When I insisted that they could, they would direct me elsewhere. While it is difficult to capture such an awkward encounter in text, I have reproduced key moments of the discussion below.

KATE: I wanted to know how it was decided which organizations would be checked?

CLERK: Any organization that collects money from abroad and that participates in political activity is deemed a foreign agent.

KATE: Yes, I know, but I want to know how you decided which organizations were to be checked, because not all nonprofit organizations in Irkutsk were checked.

CLERK: You can go to the federal prosecutor's website and there you can see a list of those organizations that have been designated as foreign agents. They are all posted online.

KATE: I don't need to know which were designated foreign agents, but rather how you decided which ones to check.

CLERK: We are not at liberty to discuss the particulars of any organization. We can only discuss information about an organization with that organization itself. Which organization do you represent?

KATE: I don't represent any organization. I don't want to know about any particular organization. I want to know about the process. How did you decide whom to check?

CLERK: You would need to talk to the Oblast prosecutor's office. It is the Oblast prosecutor that is ordering the check.

KATE: But it is the city prosecutor that is actually doing the check, right? The check is being done here in this office?

CLERK: Yes.

KATE: So then you had to have a process and you would have to know what that process was. I just want to know the process of how you decided who would be checked first, who second.

At this point, one of the women grew flustered. "I'm telling you, you would need to ask the Oblast prosecutor. All we do is fulfill the orders of the Oblast prosecutor. We received a list of organizations and we checked them." And I quickly jotted in my notes that there was a list provided to the prosecutor of which organizations to check. But the origin of the list remained obscure.

KATE: Some activists are suggesting that the checks are illegal because the documents they had to submit are from times before the law was in effect. How would you answer them?

CLERK: The job of the prosecutor is to fulfill the Constitution of the Russian Federation, which is the law. So whatever the prosecutor does is fulfilling the Constitution, so it is legal.

And so the dance continued with questions and deflections for some time. Usually, the two women talked to me pedantically, as though everything were a legal matter far above my comprehension, or with some annoyance, hinting that my questions were stupid and a waste of their time. Suddenly, the woman who had been aggressive switched personalities and began to speak to me in a sweet and conciliatory voice.

"There is nothing wrong with being a foreign agent," she said, smiling and talking gently, as though to a frightened child. "It is just a legal term. It isn't bad. No one is saying that it is wrong to be a foreign agent, that organizations shouldn't do it. It is just one legal designation as opposed to another. There is nothing wrong with registering as a foreign agent."

I left the prosecutor's office in a daze. I turned my steps toward the river, trying to process the conversation that had just taken place. I walked, lost in thought, along a street named Dzherzhinsky,[6] which, at that moment, seemed perfectly appropriate.

Legal Nihilism

Corruption has long been a part of Russian bureaucracy. Rule of law was not an accepted tenet of the Imperial autocracy (Hosking 2001), and systems of exchanged favors greased the gears of Soviet society throughout the regime's life (Ledeneva 1999). In the 1990s, a time of near total institutional collapse, corruption blossomed into a scale and form unprecedented in Russian history. Bribery, graft, and kickbacks became the accepted norm. The law was, as one Russian lawyer once told me, "artificial," because there was always a means to meet your desired end. Enormous fortunes sprang out of lawless activity, and such actions often went unpunished. Foreign companies seeking to do business in Russia have complained or even withdrawn in the face of stifling corruption (Meyer 2011).

After succeeding Putin to the Russian presidency in 2008, Dmitri Medvedev continued an anti-corruption campaign that he began in his days as first vice president. He vehemently railed against the problem of "legal nihilism" in Russia, by which he meant the epidemic of corruption and bribery in the state bureaucracy. Taking up the issue, legal scholars have examined Russians' willingness to disobey laws to measure the mood of "legal nihilism" in the country (e.g., Hendley 2012).

"Nihilism" is a strong word. As a philosophy, nihilism suggests that life is inherently without any value, object, or purpose. "Legal nihilism," then, would be belief that the rule of law has no inherent value, object, or purpose. Willingness to obey the law cannot, then, be the correct definition of "legal nihilism." For individuals on the receiving end of law's authority, their action constitutes a choice—to follow or not to follow the law. Individual disobedience does not offer us any obvious answer to the question of whether that same individual values the rule of law. By that definition, anyone with a speeding ticket or any activist committing civil disobedience would be considered a legal nihilist.

More appropriate for the term "legal nihilism" is the *production* of law. Are laws produced with the intent that they govern as the highest authority? To what extent is the law of the land expected to be universally applicable? The value of the rule of law is first and foremost found in the creation of law and its expected intent. In the case of the law on foreign agents, it is manifestly clear that the law was created to serve as a weapon of the state against its opponents. Neither was the law executed by the state in accordance with universal applicability.

I define "legal nihilism" as the politicized application of the law. It is a state practice whereby the rule of law is openly avowed, but law itself is, either by design or by execution, applied selectively in order to target and suppress certain individuals or groups that have been deemed threatening to the interests of political elites. An example of legally nihilistic design can be found in the wording of legislation. Vaguely worded texts or broadly defined legal categories are particularly useful for the practice of legal nihilism, as they open the door for selective use. When laws are both broad and vague, there is little chance that the government would actively pursue all deviants, but that most individuals or groups could be found to have violated some aspect of it, should the need arise. States may also practice legal nihilism when arbitrarily indulgent requirements are appended to the execution of a law to ensure the target's noncompliance. As the *proverka* continued, it became increasingly apparent (had there been any doubt before) that the Foreign Agent law, in design and in execution, was less the law of the land than it was a legal weapon to target state opponents.

The *proverka* was accompanied by a great deal of uncertainty, confusion, and rumor among NGOs. One question loomed especially large: Why were some organizations checked and others not? The letter of the law states that if an organization received funds from abroad and took part in political activity, it would be considered a foreign agent, but many Russian NGOs who receive foreign grants were not checked, and the definition of political activity was sufficiently vague to warrant confusion both by those who were checked and those who were not.

In April 2013, I traveled to Ulan-Ude, the capital of the Republic of Buryatia, which borders the eastern shore of Lake Baikal. Ulan-Ude is also home to environmental organizations and initiatives, and these frequently cooperate with groups in Irkutsk. There I spoke to Yegor, a man with long experience in environmental activism around Baikal.

While we were speaking, the subject of "foreign agents" came up without my prompting, and I asked whether his organization had been affected by the checks.

"No, no one has made any such mention of [my organization]," he answered. "Of course, if you look at the law and what makes foreign agents—getting foreign funding and political activity and such—then we meet all those requirements. We are the perfect candidate for such a law, but no one has said anything to us. Which is why I think that they didn't just look at all organizations that fit the categories, but instead had certain organizations in mind who for whatever reason they found bothersome, and they targeted those."

Reviving Siberian Land was checked and received an official warning. While not prosecuted as a "foreign agent," the organization was told that it was "at risk" should it continue to conduct its affairs as it has. The rationale, Tvorogova explained, was that they held a workshop and invited local representatives to attend, and that it receives money from En+.

"What?!" I exclaimed when I heard the news. For the moment, I bracketed the absurdity of a nonprofit organization not being allowed to invite legislators to a workshop and focused instead upon the surprising news of their alleged foreign funding. "But En+ is a Russian corporation!"

"Yes, it is a Russian business, owned by a Russian oligarch, and its headquarters are in Moscow," she said. "But like all big businesses, their bank account is in some island somewhere. The prosecutor just sees the black and white." Here she acted the part of the prosecutor, pretending to read documents with squinted eyes and a sour face. "They say, 'Registered in such-and-such island.' So, we have this situation where a Russian nonprofit can be troubled for taking money from a major Russian corporation."

But if association with En+ were the source of their trouble, it raised a new quandary. Every environmental organization in Irkutsk *except* for Baikal Environmental Wave received funding from En+, and yet only the Wave and Reviving Siberian Land were subject to the checks. Where were the others?

Not only was GBT among the most visible of En+'s partner organizations, it also receives support from many of the same foreign channels as the Wave—particularly since both organizations have received assistance from Gary Cook (see Chapters 2 and 3). Money from the US Forest Service supported one of the main winter projects that GBT undertook in 2012, developing brochures and interpretive materials for two of its most popular trails. It also has received funds from Pacific Environment and GlobalGreen Grants. Were one to consider the donations that foreign volunteers make, which are bundled into the cost of a summer trip, then GBT receives most of its operating expenses from abroad.

One afternoon, shortly after the Wave had received its first notice from the prosecutor, I had lunch with Marat from GBT and I mentioned the *proverka*.

"What do you think," I asked him, "will GBT be checked?"

"No," he said. "GBT is not that kind of organization."

"What kind?" I asked.

"The Wave is a protest organization," he said. "They are always against something, against some development. GBT . . . I don't know how to say it. We aren't really *against* anything."

Marat was not alone in his assessment. In Irkutsk, the Wave had a reputation as a rabble-rouser. GBT was generally perceived as kind, fun-loving, outdoorsy volunteers, largely young people, doing work projects in the spirit of the Soviet *druzhiny*.[7] However, had the prosecutor's office actually audited the organization—as it did Baikal Environmental Wave and Reviving Siberian Land—there would have been ample evidence to brand GBT as participating in "political activity."

GBT has, as an organization, signed petitions and published open letters on political topics, including closing the paper mill, rerouting the planned oil pipeline, and favoring the creation of new regional protected territories. GBT is also on an official list of nonprofit organizations who may participate as organizations in political campaigns. They have been involved in conferences and workshops with government officials. Had the prosecutor's *proverka* been nondiscriminatory, GBT could have been easily caught in a dragnet. Instead, there is reason to suggest that the law was designed to target particular nonprofit organizations who had been especially meddlesome in the affairs of the state and business elites.

The only organization in Irkutsk that was determined to be a "foreign agent" was Baikal Environmental Wave. The two other organizations that were checked and received official "warnings" were their close collaborators and associates: Reviving Siberian Land and the Center for Independent Social Research. Baikal Environmental Wave has long been a thorn in the side of the state and corporate interests, and this was not their first brush with repressive use of the law. As the frontrunner and coordinator of protest actions against the paper mill, planned oil and gas pipelines, and other threats to Baikal and its watershed, the Wave has not been endearing itself to the state. Given the Russian government's legal nihilism, it is unsurprising that the Wave is also no stranger to state suppression.

Since it took on protest activities in addition to its educational work in 2000, the Wave has been on the radar of powerful interests. During the protest against the Yukos pipeline, Jennie Sutton came home to find her apartment ransacked. Several days later, her car was stolen. While it is not known that these were done by an agent of the government, Sutton believes they were, and similar tactics have been reported by other dissidents in Russia (e.g., Henry 2010, Amos 2012; Lally 2012; Loiko 2012; Englund 2013). Around the same time period, the organization's bank account was illegally frozen. The tax office sent a letter requesting the bank to suspend the account, and the bank briefly complied, despite the fact that the letter bore no signature and was technically invalid. No valid letter ever arrived, as the organization was fully tax compliant.

In 2002, the FSB raided the office and confiscated the Wave's computers and its Internet server. The pretext on this occasion was the claim that the Wave possessed classified maps. The organization had hired two geologists to help them create maps of radioactive pollution in the Angarsk region. Baikal Environmental Wave insists that the information used was already available to the public, but their computers and Internet server were taken all the same.

The Wave's computers were confiscated again in 2010, this time under suspicion of copyright infringement. The police alleged that the organization was using pirated copies of Microsoft Office. The Wave staff produced the boxes for their software, including the sticker affirming it was an official version, legally purchased. But the police again confiscated their computers for a period of six months. Jennie Sutton tells the story thus:

> JENNIE SUTTON: There was one police officer from the Extremism division, and three from the commercial police. What was the officer from the extremism bureau doing there, if this was about Microsoft? We even had the boxes showing the software was purchased legally. And you could tell that the three from the Commercial division were uncomfortable. They knew that what they were doing was illegal. But the one from the Extremism division seemed to be in charge and was telling them what to do. They sent the case to the police in the Sverdlovskii district. We went to the police station and they gave us a good moral scolding about copyright infringement, and then we showed them that we had bought our software legally and that this was all cooked up. And after that, [the chief of police] was on our side! The prosecutor was working for the higher ups and he said, "We have to continue this case," and the police chief said, "You have no case. I've already given them back their computers." Which is why I think, at certain moments, everything depends on the individual and how an individual decides to behave in a certain moment.

The Wave had survived these persecutions in the past, but none of these was so existential a threat as the Foreign Agent law. It may be applied selectively, but once applied, it became difficult to escape the looming repercussions.

Regimes of legal nihilism may not abide by the rule of law, but neither is it rule *without* law. Once the Wave was determined by the prosecutor to be a foreign agent, the organization began a lengthy process of legal battles. Members were constantly consulting lawyers and seeking for legal precedent in crafting their defense. They had successfully reopened their closed bank account and reclaimed their taken computers by insisting on appropriate legal behavior. It was up to the lawyers and judges in Russia's widely mistrusted legal system[8] to determine the future of the Wave.

At the same time, the Wave faced additional difficulties beyond the most basic question of whether or not it was, in fact, a foreign agent. Once the prosecutor's original decision had been made, the Wave held two meetings—a staff meeting and a general assembly for all the membership—to discuss options. At both meetings, members affirmed that registering as a "foreign agent" was off the table. The group decided that they would "fight till the end," as Artur stated, to give the Wave "a beautiful finish," as another long-term member put it. Having made this choice, adherents faced the very real possibility that Baikal Environmental Wave, one of the oldest environmental advocacy organizations in Siberia, would be forced to close. This potentiality led to another legal problem, one which had immediate implications: What to do with the Wave's property.

Baikal Environmental Wave received bountiful support in the mid-1990s from the Heinrich Böll Foundation of Germany, which enabled it to purchase a spacious, two-story office in the Akademgorodok section of Irkutsk. As real estate and rent skyrocketed in the city during the 2000s, the Wave was able to stay afloat with its permanent residence intact. Moreover, it was able to rent office space on the second floor for added income in lean times. In many ways, the Wave office was its life support. Should the Wave refuse to register as a foreign agent, this property would undoubtedly be lost to it.

"I'm thinking of starting a new organization," Jennie announced one day over tea, about a week after the prosecutor's letter. I raised my eyebrows enquiringly. "I already know what I'm going to call it," she said, with a sparkle in her bright blue eyes that suggested something mischievous.

"What will you call it?" I asked, taking the bait.

"*Vtoraya Volna!*" she said: the Second Wave.

The prospect of a new organization was brought up at the Wave staff meeting as they considered their options in light of the prosecutor's decision. There seemed to be much enthusiasm for the possibility. The new organization could write a charter that would be less susceptible to prosecution for political activity, or it could strive to possibly eschew foreign funding. There was even some discussion of creating a commercial organization, rather than another NGO, that was geared toward micro-credit for environmental businesses. Marina Rikhvanova was the most excited about getting into the business of environmental entrepreneurship.

"But where would we get the money for financing loans?" Jennie asked.

"From foreign grants," Marina answered.

"Then we would be a foreign agent again," Jennie replied.

"But we would be a business, not a nonprofit."

"So businesses can get money from abroad and be involved in politics, but social organizations cannot?" Jennie asked incredulously.

"Yes," Marina and Masha answered simultaneously. Jennie dropped her pencil.

"Arghh!" she exclaimed shaking her two small clenched fists in frustration.

The principle question revolved around whether the Wave could form a new organization and transfer its property to that new organization so as not to lose what had been the Wave's greatest financial asset. At first, members were counting days—what was the maximum amount of time that it would take to officially register a new organization, and what was the minimum amount of time that the Wave could extend its legal battle in court before the property would be confiscated. Could such a property transfer be possible given the legally stated time frames for each process?[9] Despite the evident disregard for the rule of law in the *proverka*, law still defines the rules, and organizations are constrained in their own actions by legal requirements.

Back in the USSR

For those contemporary activists who remember the Soviet government, the new foreign agent law felt frighteningly familiar. As one person put it:

> In the Soviet Union there was no independence. Just like what they are trying to do today. They are trying to get their arms around social organizations again. In the Soviet Union, there were no independent organizations. Then in the1990s they started to appear and this was helped with financing from the West. So now you have government-sponsored organizations and independent organizations. This law is aimed at bringing those independent organizations back into the fold.

Another organization that went through the prosecutor's *proverka* and emerged with a warning was a research center that produces reports for scholars, businesses, nonprofits, and others who commission their work. The organization also conducts some local projects on its own, organizes conferences, and maintains an active research profile. Baikal Environmental Wave is among their regular clients. When I discussed the *proverka* with one of their lead researchers, he was also reminded of the Soviet period. "I think [the *proverka*] is a prophylactic [measure]. The result will be that social organizations will be more careful about what they do and say regarding the powers that be."

In October 2015, Baikal Environmental Wave celebrated its twenty-fifth anniversary as an organization. The following month, the Ministry of Justice of the Russian Federation for the Irkutsk Oblast found Baikal Environmental Wave to be a foreign agent and in violation of its requirement to register. On January 29, 2016, the Wave was fined 150,000 rubles, and each of its three co-directors was fined 50,000 rubles (Makhnyova 2016). An online fundraising campaign began, and within days the entire sum had been raised through local private donations. One anonymous donor gave half of the entire sum. In a prepared statement, the directors announced that they could not continue environmental work under the designation "foreign agent" for reasons that were both technical and moral. On February 3, 2016, Baikal Environmental Wave was legally liquidated by its own request.

Conclusion

The state has the privilege of writing the rules for all social fields in its sovereign territory. This right gives the state an enormous advantage over other generalizable power holders, whose influence is less explicit and less direct. While the state could use its power to directly govern the activities of its opponents, legal nihilism allows the law to be used selectively against particular groups and thus contain the reach of civil and financial power.

This tactic differs from that of the Communist Party in the Soviet Union (see Chapter 2). Recognizing that there were multiple, competing generalizable powers at play, the Communists sought to monopolize them all. The Party was simultaneously the state, the economy, the religion, and civil society. Alternative forms of each were forbidden. The attempt to monopolize the field of power was quixotic and never completely successful. It placed a heavy burden on the state, created inefficiencies, and compounded the Soviet Union's many problems. The contemporary Russian political elite, personified in Vladimir Putin, have found a new and better solution. Through legal nihilism, the Russian state has streamlined the approach to field dominance, shedding the cumbersome apparatus of the Soviet Union for a leaner, more nimble approach. Rather than monopolize all generalizable power, legal nihilism can simply limit generalizability.

Essentially, economic and civil powers are allowed to flourish independently and act on their own volition right up to the point that they infringe upon the state and its agenda. At that moment they must forfeit their power or face legally repressive force. This practice was first introduced and perfected in the economic realm. Russia's oligarchs were allowed to pursue their wealth in the global capitalist marketplace, but when they dared to use that wealth against Putin, they would suddenly find themselves investigated for tax fraud or similar corporate crimes. It only took a few examples—Boris Berezovsky,

Vladimir Gusinsky, and Mikhail Khodorkovsky—to force the remaining oligarchs into line.

With the Foreign Agent law, the same tactic is used to contain civil power. Citizens are more than welcome to form social organizations. They can try to make a difference in their communities through volunteerism and good works. Civil society may exist independently and strive to create social change, right up to the moment that they threaten state power or hinder its agenda. Then comes the *proverka*. Given the vaguely worded and broadly defined requirements for "political activity" and "foreign funding," it would not be too difficult to strategically mete out punishment. As with the oligarchs, it will likely not take long for the remainder of civil society to learn where they may and may not safely exert their efforts. The Kremlin need not monopolize alternate powers—it only needs to corral them to fortify its chosen turf. If the other powers refuse to exert themselves against it, then the state dominates the field of power by default.

Legal nihilism is the means by which such a power play is accomplished. The rule of law limits the tactical use of state power, maintaining openness in the field of power. To dominate the field of power within the rule of law, a government would have to legally constrain all alternative powers, a cumbersome task that would at best be partial and at worst garner resistance. Legal curtailment was the tactic of Soviet Communists. While the Soviet regime was certainly not a paragon of legal purity, its nihilistic tendencies were but a sideshow to the enormous legal bureaucracy by which the state governed. The novelty of so-called Putinism resides in its elevation of legal nihilism to the principal mechanism of power. Through its deployment, the state can redefine the field of power so that its path alone is made smooth.

8

Conclusion

To conduct extended field work in the Russian Federation, an American scholar must receive visa sponsorship from a research institution. As I planned to embark on the present study, I sought affiliation with a major Russian university. Everything appeared to be proceeding as planned. I had a faculty collaborator and a department that had agreed to host me. The proper documentation was working its way up the appropriate bureaucratic channels. Then, one day, I received an email with distressing news: my affiliation had been denied. A colleague disclosed secretly that a university administrator had feared my research "might pose a threat to Russian national security." What was so threatening, apparently, was my proposed intention to study civil society. Fortunately, there was more than one university that could support my work, and I sought affiliation elsewhere. This time, I took a page from the VOOP playbook and proposed to study a politically non-threatening phenomenon: "environmental protection." I left for Irkutsk some months later with a visa in hand.

Civil society is sufficiently threatening to the authoritarian state that visa affiliations can be denied because of it. It holds a type of power that the state cannot wield: the free, self-directed, and voluntary exertion of human capacity for a cause deemed worthy. This "people power"—or, as I have been calling it, civil power—may be used to produce change anywhere, in all the myriad social fields that make up human society. As a power source, it is generalizable.

Power is not uniform. Its bearers do not exert it in uniform ways. Civil society commands one type of power, and it necessarily confronts bearers of other types of power as they act in society to put forth their diverse social visions. Therefore, conceptions of civil society as external savior or bulwark against overweening power are not exactly accurate. But neither is the assumption that it is a tool of elites. It is one player among others, all of whom have their own agenda, and each of whom is limited by the capacity of the other powers as they interact in a meta-field.

The field of power is a space of dynamism and contingency, structured by the generalizable powers at play: law, money, and human beings. The ability to wield each of these arsenals may wax and wane, and players in the field may grow or decline in influence relative to one another. Power in the field is always contextual and contingent. Fissures and fractures are perpetually present, and these can be seized and utilized to alter societal trajectories to greater or lesser degrees. Recognition of the field of power helps us to better understand civil society and its position in globalized, authoritarian regimes, like Russia.

Civil Society and Authoritarianism in the Globalized Twenty-First Century

The collapse of the Soviet Union promised the "end of history" (Fukuyama 1992) and global consensus on the liberal democratic form of government and the dominance of market capitalism. But state power did not go gently—to the contrary, it did not go at all. Those governments that know the possibility of unrestricted state power have not simply surrendered it to appease Western convention. The twenty-first century remains an era of authoritarianism. These authoritarian states around the world—from China to Syria, from Venezuela to Zimbabwe—have no strong incentive to tolerate the perpetual domestic threat that civil power can pose.

But authoritarian states in the twenty-first century are necessarily different from their twentieth-century counterparts. Their power is also contextual, and that context is now global. Actors in civil society have leapt to take advantage of the global terrain. They recruit resources from wealthy nations, mobilize conscious constituents abroad, form coalitions of solidarity, and make other people's politics their own governments' concern. Activists across borders learn from one another at the collaborative World Social Forum. They travel to protest at the meetings of global governance institutions. They cooperate to confront common enemies. Scholars studying transnational activism and global civil society have often met these efforts with unabashed optimism.

It is important to remember, though, that members of civil society are not the only parties globalizing, strategizing, and learning; so too are political and economic elites, who sometimes stand in opposition to the aims of global civil society. Economic elites learn from one another how to court and co-opt their opponents in civil society, creating new norms and swapping stories in such venues as the World Economic Forum. Neither do national governments stand idly by while their domestic dissidents build their arsenals abroad. They learn and adapt, striving to preempt challenge to their dictates by legally limiting global connectivity. Every step forward taken by civil society can be met by its contenders in the field of power.

But does that mean that globalization as a phenomenon adds no unique value to the field of power? The experience of Baikal environmentalists suggests otherwise. Civil society makes use of the power of human bodies and minds, and it is precisely in the mind and its imagined horizons that the most fundamental effect of globalization can be found. The influence of global connectivity on imagined possibilities offers civil society a unique advantage in the field of power. And the possibility of imagining alternatives is a key precursor to any action that effects social change. Political and economic elites can work to constrain or channel this expansiveness, but it cannot be wholly undone. It is the Trojan horse in their midst, with no sure prediction of when its hidden fighters will burst forth.

Civil society can find assistance in this dynamic struggle with the state from their erstwhile opponents in the field of power. Governments in Russia and China embrace the market even while limiting freedom in the political realm. But economic elites have an interest in fostering civil society for their own ends, as the En+ Group does through its sponsorship of Baikal environmentalism. Economic elites require a strong independent civil society if they are to glean civil power for their own marketing requirements. Although economic elites may simultaneously conduct value projects that shore up their own legitimacy, their partnership with nonprofits promotes and strengthens civil society as a phenomenon within Russia nevertheless. Putin has made it clear that the oligarchs are not to play politics, and their work with Russian civil society organizations gives politics a wide berth, but they are maintaining civil society organizations and increasing their capacity to mobilize civil power.[1] Currently that power is not targeting the authoritarian state, but it could.

Finally, history also acts in the field of power as a structuring force and as a condition against which the future is projected. The long shadow of Communist Party rule can be seen in the fatalism and powerlessness expressed by the villagers of Bolshoye Goloustnoye. Memory of it also shaped the radical democratic organizational structure of Baikal Environmental Wave. Historical events necessarily imprint the field of power, and no analysis is complete without an understanding of historical influence on present-day action. And for that reason, although authoritarian states may seek to manage globalization, its impact prior to the erection of legal barriers can survive in collective memory. The experience of remembered freedom in the heady days of civil power will leave its own legacy. This collective memory is certainly weaker than that of state domination, which is the historical norm in Russia, and it is unfortunately tied to collective memory of collapse and crisis. But for a select subculture, that experience was formative, and it will likely continue to shape the future field of power in subtle ways, even as those days retreat into history.

The Field of Power under Putin

A closed field can last for a long time; once one player comes to dominate the field, it can be difficult for its opponents to gain sufficient strength to make more than minor adjustments to the course of development. During Soviet monopoly over the field of power, civil power could raise awareness and even win concessions from the state, but it could not prevent or redirect the military-industrial requirements of the dominant Communist Party (see Chapter 2). A field that is overwhelmingly dominated by one player has a tendency to perpetuate social problems. This tendency is especially worrisome for the environment in our modern era, where human productive and consumptive capacity have exploded to such a degree that environmental harm now accumulates with distressing rapidity (Beck 1992).

Civil society can do more to shift course in the face of mounting social problems than either of the competing generalizable powers. Civil society's generalizable power source—human beings—have direct experience with the results of these unilateral agendas, more so than absentee owners with their financial capital or political decision makers in a cosmopolitan capital. When civil power is relatively strong in the field of power, the people who comprise it can take personal and immediate action to correct social problems wherever they are found (Alexander 2006). Without this ability—whether because of state repression or limited social imagination—social ills will accumulate to disastrous effect.

The Russian state under Putin seeks to dominate the field of power, but unlike the Communist Party, it allows alternative powers to grow and flourish—as long as they do not threaten its own power and agenda. In the short run, this tactic has succeeded. Oligarchs reach unimagined sources of wealth and maintain a functional domestic economy but do not attempt to challenge Putin's power politically. Social groups may still organize and seek to make a difference—as long as that endeavor does not challenge the state. With this flexibility, Putin may succeed longer than his Soviet predecessors at staving off the accumulated detritus of the state's ambition.

But over the long term, the state's position is untenable. The field of power will necessarily recalibrate, either due to internal shifts or "exogenous shocks" (Fligstein and McAdam 2012). Allowing the financial and civil powers to develop and stabilize, while legal nihilism keeps the sword of Damocles hanging precariously overhead, is a risky position. In the past, the state could reassert authority during political transition precisely because crisis kept economic and civil powers at bay. If these are organized and developed, but for the threat of law, then any weakness shown by the state will be more easily seized.

Alternately, the field of power can be impacted from the outside. Putin's sovereignty stops at Russia's borders, but the field of power does not. Globalization is a potent force in the modern era, and despite the ideological battle that the Kremlin has undertaken against the West, the global plane will still impact Russia's domestic field of power in myriad ways.

Finally, there is a material world that supersedes the visions and agendas of any human power. In the case of environmentalism, the vision and agenda of Russia's political and even economic powers may result in the kind of ecological damage that will itself disrupt the field of power. What that kind of field-altering change in the environment might be is frightening to imagine.

Two Lessons for the West

Western governments have made no attempt to hide their enthusiasm for building civil society in authoritarian regimes. Aid and expertise are regularly channeled to activist groups abroad in the name of promoting democracy and protecting human rights.[2] The Foreign Agent law in Russia, along with similar laws that have been passed or proposed in regimes from China to Zimbabwe, shows that authoritarian governments are (rightfully) wary of this practice, which they perceive as a threat to their sovereignty, at least partly perpetrated by a foreign government in the cloak of domestic dissent. In the face of this alleged infiltration, authoritarian regimes have deemed "civil society" itself suspect. To the extent that Western zeal for building civil society has created a climate of distrust in foreign governments toward its domestic dissenters, this strategy of nation building has backfired.

The findings of this book suggest that, should the political climate prevent Western governments from directly supporting citizens' groups in authoritarian regimes, the cause is not lost. Certainly financial resources play a role; they keep groups afloat that might otherwise not survive. But the most fundamental "resource" that foreigners can offer to their counterparts in authoritarian regimes is *interaction*. Western governments need not sponsor political causes to achieve political ends. Transnational exchange is a worthy end in itself because it can expand the imagination; it can provide the space necessary to perceive the possible—which in itself can have transformative effects. Finding opportunities for transnational cooperation and exchange in any capacity has the potential for wide social impact.

Without directly funding foreign organizations, Western states and foundations can pay their own citizens to travel abroad to interact with groups in repressive regimes. They can also sponsor individuals from other countries to attend programs in their own countries. Youth camps, artist collaboratives, connections between gardening societies, and even international tourism all have the chance to spur the creative thought and newly imagined alternatives in a society geared toward fatalism and a stagnant power structure. After perceiving different horizons of possibility, individuals in authoritarian regimes can work within their unique context, as they know best, to try to put in place their newly imagined vision. The result may not resemble the democratic West, or be what Western governments would have created themselves, but they would be homegrown and domestically directed, circumventing any alleged interference from a meddling foreign state, as depicted in the Foreign Agent law.

But there is second lesson for the West that is embedded in the story of Soviet environmentalism, a warning directed to Western societies themselves. Soviet ecocide was principally the result of suppression of civil society when the Communist Party overwhelmed the field of power. The Soviet regime prevented people from acting on their own initiative to address the problems that stood before their own eyes, and it suppressed those individuals who spoke up. The result was the assumption that individual effort made no difference, a social affliction endemic to authoritarian governments, as well as any context where the field of power is highly constrained.

While the United States and Western Europe can claim a robust civil sector with high levels of engagement and few restrictions on public association and organizing, there are important implications in the Soviet experience which may be harder for the freedom-loving West to swallow. As discussed above, global capitalism has an important role to play in breaking down the barriers of authoritarian regimes. But because the field of power is *contextual*, the positive role of capitalist enterprise in one geographic and social environment can transform to a negative one in another. Such is the case in the West, particularly in the United States. There is a popular political philosophy that suggests that markets are the guarantors of freedom from a tyrannical state (Friedman 2002, Hayek 2007). But the field of power does not discriminate in its allocation of heroes and villains. No wielder of power is inherently a guarantor of any particular social good. Markets may serve to counter tyranny when the power of the state is overly strong. But in other contexts, financial power could have a monopolizing effect. The question is whether the public—who are the living arsenal of civil society—believe in their own efficacy and can imagine alternative visions for the future.

Overbearing state power created fatalism in the Soviet Union, but there is reason to suspect that overwhelming economic power would have a similar effect in the United States. Wealth inequality has reached troubling proportions in the United States (Saez and Zucman 2014), and the moneyed arsenal has shown itself more and more capable of laying claim to state power.[3] In the face of this growing concentration of financial power and its apparent influence over the state, the general public may find their sense of possibility constrained. They may slip into Soviet-style fatalism themselves. Should this happen, not only would democracy suffer, but so would all the multifaceted issues, concerns, and causes toward which civil society turns its attention every day

Figure 8.1 A GBT volunteer gazes at the Idol, one of several named rock formations accessible by hiking trail in the Baikal basin. She wears a shirt that she received from REI while volunteering with Earth Corps, a youth environmental exchange program in Seattle. Photo Credit: Nancy Hollman

(Fig 8.1). One need only look at the plight of nature in the Soviet Union, from the Aral Sea to Chernobyl', to recognize the fearful consequences that occur when the public cannot redirect the course of action set by a concentrated power.

Baikal Forever

One brisk day in early April 2013, my husband and I joined three friends and a small dog for a trek across the frozen surface of Lake Baikal. Starting at dawn, we walked from the southwestern bank to the southeastern shore, with occasional breaks to warm ourselves with snacks and hot tea from a thermos. One of our companions, Katya, was preparing to leave for a six-month volunteer program in Latvia. She had previously spent a year and a half doing environmental volunteer work in the United States, and she was romantically involved with a young man in Spain, whom she had met on a GBT trip. I asked my peripatetic friend where she thought she might settle down: Latvia, the United States, or Spain? She laughed and demurely declined.

"I like to go abroad and volunteer, but I could never leave forever," she told me. "I love Baikal. I could never be away from Baikal for very long. Whenever I am away, I yearn for Baikal. I will always live in Irkutsk because I need to be close to Baikal."

As we trudged through the snow, with Baikal ice beneath our feet, surrounded by mountains and beneath a mercurial sky, Katya declared her allegiance to this place—to Lake Baikal. And so the field continues. Katya is one of many people around Baikal who have volunteered with GBT, participated in TBI, and learned from the Wave, or otherwise interacted with environmentalism around the Lake. They are the embodiment of civil society, and they are players in the field of power. The Russian state will strive to legally limit their efforts, and oligarchs will try to curtail their claims. Katya will keep giving of her time and treasure, freely and voluntarily, and in concert with others, for in her mind there is no cause worthier than saving the Sacred Sea.

Coda

I left Irkutsk in 2013. It has been five years since I last saw Baikal, its waves lapping against the rocky shore. In that time, the Sacred Sea and its defenders have seen one fantastic victory, but otherwise the story remains one of an ecosystem playing defense. New and ominous threats have arisen in recent years. Irkutsk environmentalists continue to keep their ears to the ground. Daily, they strive to meet the challenges that they find, both mundane and momentous.

First, the one fantastic victory: very shortly after I left the field, in June 2013, Prime Minister Dmitri Medvedev announced that BTsBK would be shuttered for good. The infamous plant, which sparked the modern environmental movement not only in Irkutsk but throughout Russia as a whole, ceased operations in October 2013. Although the plant had faced constant pressure from environmentalists, they do not take credit for its demise. Economics, not environmentalism, killed the paper mill, which was too expensive to upgrade and could not compete in the global marketplace. Still, after nearly sixty years of toxic industrial production on the shores of a pristine world treasure, its closing was met with a sigh of relief.

Unfortunately, the plant left behind six million tons of toxic sludge currently stored in tanks on the grounds of the plant complex on the shore of Baikal. The tanks are not sealed, so there is the continued possibility of groundwater contamination. Moreover, the tanks are located in a mudslide zone. In the event of a mudslide off the surrounding mountains, or an earthquake in the seismic region, these six million tons of toxic contaminants could flow into Lake Baikal. As of this writing in the fall of 2017, the government has made no decision regarding the elimination of this toxic legacy on the shore of the Sacred Sea.

Meanwhile, in 2015, residents at Lake Baikal previewed a new threat that could potentially worsen with time: water in the lake dropped to its lowest level on record. The reason was twofold. First, the managers of the hydroelectric dam in Irkutsk took more water than usual for generating power. At the same time, lower levels of winter snowfall diminished the amount of water flowing into the lake from its 330 feeder streams and rivers. The federal government has established 456 meters as the minimum depth that is considered ecologically safe for Lake Baikal, but in 2015, the level dropped 12 centimeters below this minimum. The Irkutsk hydroelectric dam had to receive special dispensation from the government to continue powering its capital city. Villages that surround the lake saw their wells run dry. There was also concern that Baikal fish, such as the *omul*, might be unable to spawn if the water level fell too much. Because of the steep decline of the lake's shore, the fish might be unable to swim up to the shallow rivers and streams that serve as their breeding grounds.

For some, the low precipitation of 2014–2015 represents only the beginning of possible upsets to the ecological balance at Baikal wrought by a changing climate. But others point to a more immediate source for concern about lowered water levels. The Mongolian government has begun constructing hydroelectric dams on tributaries of the Selenga River, which supplies more than half of Baikal's water. Filling the new Mongolian reservoirs would dramatically lower the level of Baikal, with potentially dramatically negative impacts to its ecosystem. The World Bank, which was funding the hydroelectric projects, has suspended its support pending environmental impact and stakeholder studies. The United Nations has also gotten involved because Lake Baikal is a World Heritage Site. Mongolia insists it has a right to produce hydropower from rivers on its sovereign territory. At present, the future of the Mongolian hydroelectric dams and lake's water level are in diplomatic limbo.

Even in the Soviet days, environmentalists encouraged tourism to Lake Baikal as a potential replacement for heavy industry in order to prevent pollution, and many environmentalists since the Soviet collapse have sought to turn that vision into reality. The founding of GBT was premised upon that very idea: that the region could develop economically and sustainably through ecotourism. But their efforts to bring tourism to the region conjoined with other social forces (e.g., private business interests, state negligence, bureaucratic ineptitude, etc.) to produce some unintended negative consequences. Environmentalists' dream that Lake Baikal would attract tourism has now become a reality; unfortunately, their stipulation that it be "sustainable tourism" has not. The negative effects of unregulated tourism have compounded, and Baikal is now increasingly a victim of its own success as a tourist attraction.

The principal culprit is the sudden and overwhelming growth of the algae spirogyra on the shoreline of Baikal. The lake has always been prized for the near perfect clarity of its water, but in recent years, the shallows and beaches have been overwhelmed by a thick, green slimy mess. Spirogyra has flourished along with the nutrients from the run off of human waste and the growing use of phosphate-based detergents in new hotels and tour bases. It also thrives from the waste that is dumped into the water by tourist boats, which are in ever-greater supply. Meanwhile, Baikal's surface temperature has been increasing over the years, likely associated with global climate change, and the warming has further encouraged the explosive algae colonies. Spirogyra is choking out native endemic species, especially the Baikal sponge, which is suddenly and massively dying out around the lake. The algae infestation is associated with the general problem of litter and rubbish collection infrastructure around the lake, which is much discussed but as of yet unsolved.

Of the three organizations that principally comprised my study, and which were chosen for their strength, their dense network ties and their longevity, only one remains: the Great Baikal Trail. It continues its work. Many of the enthusiastic young people I encountered during my fieldwork are still involved, although they have also grown up and furthered their personal and professional lives. Katya has gotten married; Lusiya had a baby. Some have moved away, or moved on, while others are still with the organization.

The most important change for GBT in recent years was that it finally purchased its own, permanent office space, with money from individual donors (including me) and matching funds from En+. The office is located in the center of town and, although modest, it provides security, stability, and permanence to an organization that was accustomed to moving almost every year. It also provides protection from Irkutsk's high rental costs in lean times. And times are lean, indeed. Western sanctions imposed since Russia's 2014 annexation of Crimea have caused a spike in living costs throughout Russia. The expense of running a volunteer vacation project has likewise risen. Russians' incomes no longer go as far as they used to, and fewer people are willing to pay the cost of a GBT excursion. With such high costs, foreign participation cannot cover the majority of expenses the way it used to. This year, GBT will only conduct five summer projects, compared to eleven in previous years.

When I queried whether they had seen a decrease in foreign volunteers following the Ukrainian civil war and the annexation of Crimea, GBT reported that their fortune in recruiting foreign participants varies every year and they never really know why. They did see a drop in American volunteers in 2016 but were loath to attribute it to geopolitical tensions. In fact, when I asked about it, Dina seemed surprised that politics could impact Western tourism to Russia. "People here are still the same kind and hospitable people as before," she said. Then she added, "Well, most of us." But she also said that the drop in American volunteers was made up for by a surprising bump in participation from Switzerland.

GBT furiously writes grants for support to continue its work. It hopes against hope for governmental funding. As one informant said to me recently,

"We thought that since 2017 is officially the Year of the Environment in Russia, and it's the 100th anniversary of the creation of the *zapovednik* nature reserves, we could find some kind of state support, but nothing has changed. We have to rely on ourselves as usual." Among its fundraising initiatives, GBT opened a souvenir shop called The Wood Shop [*Lesnaya Lavka*] that sells hand-made crafts from Baikal-sourced materials and maps of the area. It is not yet profitable.

TBI had already been suffering financially for some time when I began my research. As Russia developed, the costs of the program increased, and rising tuition lowered the number of interested participants. As the Cold War retreated in the public mind, a program geared toward bringing together Russian and American youth lost its broad-based appeal for many funders. At the same time, the ambition of TBI to truly address watershed issues drove the organization to expand into Mongolia and examine the problems of the Selenga River, Baikal's largest tributary. But adding Mongolia to the mix only further increased the program's logistical burden and cost. No one knows how much longer TBI might have survived, or whether it might have discovered a new route to success, but in 2013, Bob Harris, who was the organization's most active and supportive board member, suddenly and tragically died. Without his leadership, the remaining board members were unable to rise to meet the needs of their foundering organization. In December 2013, TBI suspended its activity indefinitely.

Baikal Environmental Wave also liquidated itself in 2015, its demise due to the Foreign Agent law (see Chapter 7). The organization's headquarters passed on to an organization called the Baikal Interactive Environmental Center, which had operated under the auspices of Baikal Wave but was an independently registered nonprofit. It was this "daughter" nonprofit that ran the Eco-Schools program and conducted the "Footprint of Commodities" game at SHEPR and elsewhere. That organization continues its educational work, and some of the Wave's employees have found a new home there, from which they continue their environmental outreach work.

Other individuals associated with the Wave continue their initiatives in a variety of other capacities. Marina Rikhvanova works closely with Elena Tvorogova at Reviving Siberian Land. Among their projects, they are promoting environmental designs that turn waste and pollutants into useful products. A recent iteration of SHEPR required all entrepreneurial projects to focus upon the problem of invasive spirogyra. They also experiment with using organic waste as fertilizer. Rikhvanova is collaborating with a local Makers' Union (*Coyuz Kreatorov*) to build a factory-laboratory to test new materials made out of spirogyra and other waste products. She helped establish a branch of the Slow Food movement organization in Irkutsk. She also works with botanists and landscape designers to develop an endemic seed bank and promote the use of native plants in landscape design to preserve endemic, and increasingly endangered, Baikal flora. Neither has she abandoned advocacy. Rikhvanova and Tvorogova have also been closely monitoring the progress of the Mongolian hydroelectric dams, evaluating environmental impact studies and ensuring compliance with requisite public hearings.

Jennie Sutton also pays close attention to environmental issues around Baikal and supports the myriad projects to protect it. She still translates environmental tracts and scientific work on environmental issues to encourage the flow of ideas and knowledge between Russia and the West. Sutton maintains an off-the-grid dacha that she hopes to develop into a kind of "sustainable demonstration plot." In the meantime, the dacha offers her a regular retreat, in which she can escape the busy world and its many problems, and partake instead of the peace and quietude of nature, abundantly proffered by the generous and voluptuous Siberian *taiga*.

Notes

Chapter 1

1. Except for organization leaders, whose full names are provided, all names have been changed to protect the subjects' identities. All translations from the original Russian are the author's own.
2. "Russia—our holy homeland," is the opening line of the Russian national anthem, formally named the "State Hymn of the Russian Federation."
3. Indeed, my proposal to study "civil society" caused one state university administrator to declare me a "threat to Russian national security."
4. The so-called third sector (Anheier and Seibel 1990; Evers and Laville 2004; Viterna, Clough, and Clarke 2015), which comprises nongovernmental organizations (NGOs) and philanthropy groups.
5. I am not the first to refer to a "field of power." Bourdieu employs the same phrase sporadically in his writings (Bourdieu 1989, 1998, Bourdieu and Wacquant 1992). However, Bourdieu never systematically defines the field of power, and his usage of the term is inconsistent. This book attempts to correct that lack of specificity, taking the subject as its central focus.
6. I can imagine another generalizable power, although it does not bear on the present analysis, in the form of religion—which holds a kind of "existential" power: the legitimate interpretation of the meaning of life. In modern society, religious pluralism has removed the generalizability of religion's power source, relegating religious communities to their own particular social fields. However, in much of human history, and indeed, in some areas of the world at present, religion is more uniform and religion's power can be said to be generalizable. The interminable conflicts between Catholic popes and European monarchs can be analyzed as activities in the field of power. Similarly, the Soviet insistence upon atheism and Vladimir Putin's protection of Russian Orthodoxy are likewise power plays that involve the existential power of religion. However, religion did not impact the field of environmental protection at Lake Baikal as a generalizable power, nor do I think it could have. Despite Putin's present efforts, contemporary Russia more greatly resembles the religious pluralism of contemporary Europe than it does the Orthodox nation of the Imperial tsars, although that may change over time.
7. Readers may recognize in this discussion some elements of Foucault's (1978, 2003, 2007) concept: biopower. While Foucault recognizes the importance of mobilizing human bodies en masse to enact certain social projects, he fails to parse and differentiate the means by which these mobilizations take place. The result is a terminology creep that has allowed the phrase "governmentality" to become a catch-all, employed

haphazardly and leaving our understanding and elucidation of power still more opaque than it was before. Moreover, Foucault is principally concerned with the state, which offers a limited view of the means by which human beings can be made into the raw material of other people's social projects. "Governance" has an "everywhere and nowhere" quality that lumps rather than parses the many players involved. To understand the dynamics of power, it is necessary to pull the pieces apart and place them in a field. For my purposes, the ability to mobilize human beings is epiphenomenal to the ability to command a form of power. The type of power commanded determines the means by which "biopower" is accomplished: the law conscripts bodies; corporations must exchange money to "rent" them. Only civil society mobilizes humans consensually and with conscience. That is why only civil society can command this power—the others must deploy alternate power sources to gain access to it.

8. Although originally coined to describe Italian fascism, the word migrated to a descriptor of Nazi Germany and Stalin's Soviet Union, mostly due to the writings of Hannah Arendt (1951). However, it did not disseminate into common parlance until the advent of the Cold War, and it was synonymous in the public mind with the USSR (Geyer and Fitzpatrick 2008).

Chapter 2

1. A *poznaya* is an eating establishment that serves Buryat cuisine, particularly the regional specialty—*pozi*, which are ground beef dumplings.

2. Elderly woman; pl. *babushki*

3. Much has been written about Lake Baikal. For readers who would like to tackle the subject in more depth, here are some suggested sources: Breyfogle 2013, 2015; Bull 2001; Galazii 2012; Mathiessen 1992; National Geographic 1992; Thomson 2009; Weiner 1999, chap. 16.

4. Although the cause of Baikal's thorough oxygenation is not known for certain, most scientists agree that the intense cold of Siberian winters, which annually causes meter-thick ice to form at the lake's surface, likely plays a role. For this reason, climate change poses a significant threat to Lake Baikal and its thousands of endemic species who are uniquely adapted to its oxygenated depth (Thomson 2009).

5. Readers interested in learning more about these sporadic revolts can look at the following histories: the Pugachev rebellion (Alexander 1969), the Decembrists (Mazour 1937), the Populists (Pipes, 1964, Belfer 1978, Field 1987), and the People's Will (Offord 1986).

6. The Soviet Union ended following an attempted coup by hard-liners within the Politburo who opposed Gorbachev's reforms, specifically a renegotiation of the Union's central treaty. The coup failed when the military refused to fire upon Russia's White House with hundreds of civilian supporters gathered outside of it. Although he emerged from the coup still the nominal leader of the USSR, Gorbachev had lost his political authority. Russian president Boris Yeltsin, who emerged from the coup as a national hero, banned the Communist Party within Russia. Satellite republics announced their departure from the Soviet Union. Gorbachev dissolved the Union on December 26, 1991. For more detailed information, see Dunlop 1993.

7. *Zapovednik*, which translates roughly as "territory where it is forbidden to go," is one of several designations for protected territory in Russia. Others include national parks, state forests, game reserves and natural monuments. *Zapovednik* is the highest level of preservation and the strictest form of protection, with the aim of completely isolating ecosystems from any human interference. Under the Soviet government, the *zapovedniki* became the largest system of nationally protected lands in the world. For a thorough history of the *zapovednik* in Russia, see Weiner 1999 or Shtil'mark 2003.

8. More of the rock was visible before the construction of the hydroelectric dam in Irkutsk, which raised the water level of the Angara several meters, also submerging the train tracks that had connected the historic Circum-Baikal Railway to Irkutsk.

9. Although the acronym "BPPM" would be easier on English readers, I am retaining the Russian acronym "BTsBK" because this is how the Russian public always refers to the mill (pronounced <beh-tseh-beh-kah>) following its Russian name: *Baikal'skii Tsellyuznii i Bumazhnii Kombinat*. In the field, this acronym was referenced so often that to call it by any other name would, to my mind, be inappropriate.

10. However, the efficacy and proper usage of these measures were debatable. Needless to say, for the environmentalists, they were entirely insufficient and provided only a veneer of deniability for the regime.

11. I first encountered the claims of Soviet censorship over the subject of the Baikalsk mill while talking with informants in the field; however, I was unable to document the claim. Historian Nicholas Breyfogel explained to me the rationale behind the activists' suspicion in a personal communication.

12. At moments, the minutes of a VOOP meeting resemble exegesis from ecological modernization theory (Buttel 2000). Emanating as it does from the Soviet Union, it serves to remind us that "actually existing state socialism" is still modernity by other means.

13. And for a discussion of the outcomes of foreign funding in the 1990s, see Henderson 2003.

14. The VOOP archives contain a letter from *Vostochno-Sibirskaya Pravda* to VOOP in 1991 explaining that, since the newspaper must now earn its revenue by advertising, it can no longer provide regular space to the organization, but VOOP was welcome to purchase a full-page ad if they would like to continue to spread their message to the public (VOOP Archive 1991; Delo 735; 45–47).

15. However, it is worth noting the diffuse effect that the attention paid to environmental protection may have had on the public mind. Baikal and the importance of Baikal were repeatedly pronounced on the pages of *"Rodnik."*

16. The law may be found at http://base.garant.ru/2157025/1/#block_100.

17. For further reading on Soviet industrial pollution, see Feshbach and Friendly 1992, Komarov 1978, and Peterson 1993.

18. Declaring the country officially atheist and propagandizing atheism similarly worked to break the generalizable existential power of the Orthodox Church, a power already rendered moot in the West through religious pluralism.

Chapter 3

1. Except in so far as industrial collapse slowed the rate of pollutant emissions (Henry and Sundstrom 2007).

2. Sutton rejects the interpretation that Baikal Environmental Wave was strengthened by virtue of its "cosmopolitanism," largely because the word "cosmopolitan" carries a negative connotation in Russia. However, whether the organization would call itself cosmopolitan or not, I maintain that the attraction professed by virtually all the early members to working with international volunteers and practicing foreign language skills indicates that 1) the organization had a cosmopolitan character, and 2) that it was a powerful draw in fostering its local membership.

3. Khodorkovsky and his business partner were imprisoned by Vladimir Putin for fraud and tax evasion, and the Russian state seized Yukos and sold its assets to state-owned Rosneft at fire-sale price. It is widely believed that Khodorkovsky was imprisoned because of the political threat he openly posed to Vladimir Putin (Zolotukhina 2013). After eleven years in prison, Khodorkovsky was pardoned by Putin in December 2013, in advance of the winter Olympic Games at Sochi. He quickly relocated to Switzerland, where he is now a permanent resident.

4. Sutton's activism against the Yukos pipeline and additional pipelines that were then proposed by Transneft and Kovytka kept her in the forefront until 2005. On other occasions, Sutton cited different reasons for her resignation as co-leader of Baikal Environmental Wave, and her concern for the image problem created by her "foreignness" was only one among several considerations.

5. The Yukos pipeline out of Angarsk was finally prevented by a representative of the State Committee on Environmental Protection who refused to sign off on the project's environmental impact report—a requirement for any new development at the time. She was dismissed from her post within a few months, leading local environmentalists to suspect retaliation by the power elite.

6. According to some tellings, President Reagan himself called Deukmejian and recruited him to the project.

7. TBI participants were not uniformly American across the years. The SEE accepted applicants from other countries as well, although all participants were required to speak either English or Russian. However, the Russian participants (with one exception) were always drawn from the Baikal region.

8. She went on to say: "I still have very, very, very warm feelings in relation to everyone in the (US) board of directors. For Bob Harris, and Charles Goldman, and John Gussman, and Karen Fink, whom I really love . . . In general I don't have any hard feelings for anyone."

9. The Russian name for the organization is *Bol'shaya Baikalskaya Tropa*—literally, the Big Baikal Trail—but I follow the English name by which the organization refers to itself.

10. The environmental brigades branch of the Komsomol (Communist Youth League).

11. The actual quote, from *The 18th Brumaire of Louis Bonaparte*, reads, "Men make their own history, but they do not make it just as they please."

Chapter 4

1. As of this writing, the Tahoe-Baikal Institute has suspended its operations, after nearly a quarter-century of coordinating its environmental education and cultural exchange schools. The invisible tie that binds Baikal and Tahoe has become more tenuous with its passing.

2. "Bolshoye" and "maloye" are adjectives that translate as "big" and "small," respectively. They designate two villages: Big Goloustnoye and Small Goloustnoye. Curiously, Maloye Goloustnoye is larger in population than Bolshoye Goloustnoye (1200 and 600 people, respectively). The name Goloustnoye roughly translates as "naked mouth." It is presumed that the name refers to the river delta on which the two villages sit. Bolshoye Goloustnoye is situated directly beside Lake Baikal where the Goloustnoye River flows into it. Maloye Goloustnoye is located about twenty minutes upriver from Bolshoye Goloustnoye in the Pribaikalskiy mountain range outlining the western edge of the lake. The river delta is wider near the lake, hence the designation "big naked mouth" for the village near the shore, and "small naked mouth" for the village further upstream where the delta narrows and becomes smaller.

3. The SEZ planned for Bolshoye Goloustnoye was relocated to Baikalsk, to ease the economic shock to the region when the paper mill was finally closed. The Bolshoye Goloustnoye villagers may yet elude their once seemingly inevitable economic development.

4. I proceeded to tell her about "company towns" in the United States, but she dismissed the possibility that America might have had anything so similar to the Soviet system.

5. Quantitative studies find scant empirical evidence for cultural difference in political systems (e.g., Booth and Seligson 1984, Tiano 1986, Seligson and Booth 1993), but it is worth remembering that the abstraction of a Likert scale opinion query may bring its own biases, divorced as it is from context and actual observed behaviors. What citizens desire may differ from what they deem possible and worth striving for.

6. Geo-tourism is defined by the National Geographic Society as tourism that sustains or enhances the geographical character of a place—its environment, culture, aesthetics, heritage, and the well-being of its residents.

7. The ecosystems surrounding Lake Baikal vary substantially. Baikalsk has a great deal of snowfall. In the winter, it mounds up for several feet. Baikalsk is home to a mountain ski resort because of the quantity and quality of its snowfall. Bolshoye Goloustnoye, on the other hand, receives the greatest number of sunny days for any place in all of Russia. When snow does fall, it is often powdery and quickly blows away, off the dry steppe, on the strength of Baikal's wind. Ice skating is a common sport in Bolshoye Goloustnoye, but skiing and snow-shoeing are impossible there.

8. The webinar dates and topics were as follows: February 5—tourism, February 8—economic diversification, February 12—poverty, February 19—food, February 26—youth, March 5—the sister-city relationship.

9. I do not have exact numbers for attendance in all locations for each webinar. I was unable to attend one of them, and I only counted the number of participants in the location I was observing at each webinar. When I later asked Baikal Environmental Wave for the number of participants, they did not match the numbers that I had myself recorded (perhaps they included support staff), so rather than supply exact figures, I can only offer the range and the general sense of regular increase in involvement.

10. The word "*dacha*" comes from the Russian verb, "to give," because the state would give out these parcels in return for service rendered.

11. The climate in Baikalsk is uniquely beneficial for strawberries in that region, and locals have always sold the berries for supplemental income. Baikalsk has an established reputation for strawberries. In one of its many attempts to help Baikalsk find alternate revenue to replace the paper mill, Baikal Environmental Wave helped Baikalsk start an annual Strawberry Festival that is now a proud tradition, attracting visitors from Irkutsk, Ulan-Ude, and other neighboring towns.

12. In Russian, the term is "brother-cities," and the conversation began with the group musing on the culturally distinct gendering of inanimate objects, such as cities.

Chapter 5

1. Oleg Deripaska's personal website, accessed on September 5, 2013: http://www.deripaska.com/in_focus/detail.php?ELEMENT_ID=691#.UikuSj_fIug.

2. Oleg Deripaska's personal website, accessed on September 5, 2013: http://www.deripaska.com/in_focus/lets_discuss/knigi-kotorie-ya-chitayu/#.UikN2T_fIug.

3. Unless otherwise noted, the information provided about the En+ Group comes from the company website: http://www.enplus.ru/about/.

4. This vision for Siberia is not new. Soviet central planners likewise viewed Siberia as a source of future development and growth (Roe 2016).

5. Where not otherwise referenced, the information provided on Deripaska's biography can be found on his personal website: http://www.deripaska.com/.

6. Fourteenth richest as of 2012.

7. For an interesting personal reflection on the various forces at play in Russia's economic transition, see Jeffrey Sachs's (2012) essay, "What I Did in Russia." http://jeffsachs.org/2012/03/what-i-did-in-russia/.

8. I did not observe great enthusiasm among the recruited participants nor did I hear expressions of how excited they would be to participate again.

Chapter 6

1. The first wave was preservation/conservation; the second involved end-of-pipe pollution and command-and-control regulation. Third-wave environmentalism focused upon environmental racism and environmental justice.

2. A classic comparison is the alternate responses to the ozone hole and acid rain. Regulation, albeit with strong industry buy-in, effectively ended trade in CFCs and has largely solved the problem of the ozone hole. The market-based cap-and-trade approach to acid rain slowed the rate of acidification in forests, but it has not prevented its continuation.

3. The actual organization name in Russian is *Vozrozhdenie Zemli Sibirskoi*, which literally translates as the Renaissance of the Siberian Land. I thought this translation cumbersome, and that Reviving Siberian Land was more apt for their intentions, catchier as an organization name, and easier on the tongue for English speakers—just as the name *Vozrozhdenie Zemli Sibirskoi* is for Russian speakers. Still, I have seen the organization's name written in English elsewhere according to the more literal translation. In case one is looking for the organization in other sources or venues, I wanted to be clear that Reviving Siberian Land and the Renaissance of the Siberian Land are, in fact, the same organization.

4. Literally, "Eco-Footprint of a Commodity," but I changed it to the plural form for the sake of stylistics.

5. In statistical analysis, degrees of freedom represent the range of possible states in a sample, determined by the number of independent variables. I borrow the phrase here to suggest that socialization programs hold certain elements constant, and allow flexibility in certain other elements. The greater the flexibility, the more "freedom."

6. Although "motif" might be a better word for the En+ summer camp.

7. In fact, another local environmental nonprofit, Defending Baikal Together, was also affiliated with a Russian corporation, which was its primary sponsor. However, its environmental education projects were not value projects, from my observation of them. Interestingly, it began as a bottom-up initiative, when local company employees wanted to organize a litter clean-up. As their initiative grew, the company hired a staff person to manage the activities, which then spun off as Defending Baikal Together. Although corporate affiliated, the organization did not ever use its parent-company's brand. Nothing in its program would suggest it was affiliated with business, and its environmental education projects, while socialization oriented, were not value projects, as I have defined them here.

Chapter 7

1. Examples of globally normative frames include "fighting global terror" while shoring up territory in the Caucasus (Headley 2005), "streamlining bureaucracy" when eliminating the State Committee on Environmental Protection (Henry and Douhovnikoff 2008), and "protecting children" when discriminating against homosexuals (Chan 2017).

2. There was a list of foreign foundations who were exempt from the tax, but the requirements to join this list were highly restrictive. Small sums could be renamed "charitable donations" rather than "grants," and thus circumvent the tax, but these restrictions still limited the power, strength, and growth of the nonprofit sector in Russia.

3. The text of the law can be found athttp://ntc.duma.gov.ru/duma_na/asozd/asozd_text.php?nm=121-%D4%C7&dt=2012, accessed on November 17, 2013. All translations are mine.

4. The full list of exemptions is as follows: "By political activities, not included is activity in science, art, culture, health care, prevention and protection of public health; social support and protection; protection of motherhood and childhood, and social support to the disabled; promotion of healthy lifestyles; physical culture and sports; protection of flora and fauna; charity activities, as well as activities to promote philanthropy and volunteerism" (Federalnii zakon ot 20.07.2012 g. N 121-F3).

5. Literally: "*Inostanii* [*sic*] *agent* ♥ USA" [roughly: forein agent ♥ USA]. "Foreign" was misspelled, and while "foreign agent" was written in Russian Cyrillic letters, "USA" was written in Latin lettering, as opposed to the Cyrillic translation: "СIIIА."

6. Felix Dzherzhinsky was the first director of the Soviet secret police. Originally called the Cheka, it is better known by its later name, the KGB.
7. The environmental brigades of the Komsomol [Communist Youth League].
8. Numerous polls have reported low levels of public trust in Russia's legal system. A few recent examples include: Cheloukhine, Ivković, Haq, and Haberfeld 2015; Kozyreva and Smirnov 2015; Trochev 2016.
9. The office was successfully "gifted" to a spin-off organization dedicated to environmental education.

Chapter 8

1. Moreover, the stable and growing economy that is the oligarchs' chief concern is building a solid middle class that, in the right historical moment, could provide civil society the necessary force to make a successful power play against the state. Such an opportunity is not a given, but it is more possible with the presence of a strong middle class than without it.
2. Indeed, the three principle organizations comprising this study have all received funding from the US government at some point.
3. The US Supreme Court decision *Citizens United v. FEC* has significantly altered the field of power and the relationship between state and financial power holders. While this was perhaps the most obvious and consequential power play, it symbolizes numerous other such plays that more regularly take place through donations, gifts, and the "revolving door" between the political and economic institutions in the United States.

Bibliography

Ahmed, Kamal. 2012. "Oligarch Oleg Deripaska Reveals He Paid Armed Gangs to Protect His Business Empire." *The Telegraph*, April 21.

Alexander, John T. 1969. *Autocratic Politics in a National Crisis: The Imperial Russian Government and Pugachev's revolt, 1773–1775*. Bloomington: Indiana University Press.

Alexander, Jeffrey. 2006. *The Civil Sphere*. New York: Oxford University Press.

———. 2013. *The Dark Side of Modernity*. Hoboken, NJ: Wiley.

Alexandrova, Lyudmila. 2013. "Polluting Baikal Pulp and Paper Mill to Be Shut." *ITAR-TASS*, June 20.

Almond, Gabriel, and Sidney Verba. 1963. *The Civic Culture*. New York: Little, Brown.

Almond, Gabriel A. 1983. "Communism and Political Culture Theory." *Comparative Politics* 15(2): 127–138.

Amos, Howard. 2012. "Russian Police Launch Raids ahead of Anti-Putin March." *The Guardian*, June 11.

Anderson, Benedict. 1983. *Imagined Communities*. New York: Verso.

Anheier, Helmut K., and Wolfgang Seibel, eds. 1990. *The Third Sector: Comparative Studies of Nonprofit Organizations*. Vol. 21. Boston: Walter de Gruyter.

Appadurai, Arjun. 1988. "How to Make a National Cuisine: Cookbooks in Contemporary India." *Comparative Studies in Society and History* 30(1): 3–24.

Appel, Hilary. 2004. *A New Capitalist Order: Privatization and Ideology in Russia and Eastern Europe*. Pittsburgh: University of Pittsburgh Press.

Arato, Andrew. 1981. "Civil Society against the State: Poland 1980–81." *Telos* 47: 23–47.

Arendt, Hannah. 1966. *The Origins of Totalitarianism*. New York: Houghton Mifflin Harcourt.

Arutunyan, Anna. 2014. *The Putin Mystique: Inside Russia's Power Cult*. Northampton, MA: Interlink.

Associated Press. 2007. "The Political Career of Boris Yeltsin." *The Guardian*, April 23.

Babb, Sarah. 2003. "The IMF in Sociological Perspective: A Tale of Organizational Slippage." *Studies in Comparative International Development* 38(2): 3–27.

Barry, Ellen. 2009. "Putin Plays Sheriff for Cowboy Capitalists." *The New York Times*, June 4.

———. 2011. "Rally Defying Putin's Party Draws Tens of Thousands." *New York Times*, December 10.

Barry, Ellen, and Michael Schwirtz. 2012. "After Election, Putin Faces Challenges to Legitimacy." *The New York Times*, March 5.

Bartley, Tim. 2003. "Certifying Forests and Factories: States, Social Movements, and the Rise of Private Regulation in the Apparel and Forest Products Fields." *Politics and Society* 31(3): 433–464.

———. 2007. "Institutional Emergence in an Era of Globalization: The Rise of Transnational Private Regulation of Labor and Environmental Conditions." *American Journal of Sociology* 113(2): 297–351.

Bauer, Raymond A. 1952. *The New Man in Soviet Psychology*. Cambridge, MA: Harvard University Press.

BBC. 2013a. "Russian Billionaire Gives Away $3m Bonus to Staff." *BBC Business*, July 5. Accessed on February 1, 2015. http://www.bbc.com/news/business-23195436.

BBC. 2013b. "Profile: Mikhail Khodorkovsky." *BBC News*, December 22. http://www.bbc.com/news/world-europe-12082222.

Beck, Ulrich. 1992. *Risk Society*. London: Sage.

———. 2006. *Cosmopolitan Vision*. London: Polity.

Belfer, E. 1978. "Zemlya vs. Volya: From Narodnichestvo to Marxism." *Soviet Studies* 30(3): 297–312.

Bell, Terence. 2016. "The World's Biggest Aluminum Producers." *The Balance*, August 3.

Berman, Elizabeth Popp. 2014. "Field Theories and the Move Toward the Market in US Academic Science." *Political Power and Social Theory* 27: 193–221.

Bershidsky, Leonid, Maria Rozhkova, and Maxim Tru. 2001. "Deripaska Barred from Davos." *The Moscow Times*, January 26.

Beyerlein, Kraig, and John Hipp. 2006. "A Two-Stage Model for a Two-Stage Process: How Biographical Availability Matters for Social Movement Mobilization." *Mobilization* 11(3): 299–320.

Bhabha, Homi. 1984. "Of Mimicry and Man: The Ambivalence of Colonial Discourse." *October* 28: 125–133.

Blair, Jennifer, and Florence Palpacuer. 2012. "From Varieties of Capitalism to Varieties of Activism: The Antisweatshop Movement in Comparative Perspective." *Social Problems* 59(4): 522–543.

Blam, Yuri, Lars Carlsson, and Mats-Olov Olsson. 1998. "Institutions and the Emergence of Markets—Transition in the Irkutsk Forest Sector." Interim Report IR-00-017 prepared for International Institute for Applied Systems Analysis, Laxenburg, Austria.

Bloom, Paul N., Steve Hoeffler, Kevin Lane Keller, and Carlos E. Basurto Meza. 2006. "How Social-Cause Marketing Affects Consumer Perceptions." *MIT Sloan Management Review*, Winter 2006.

Booth, John A., and Mitchell A. Seligson. 1984. "The Political Culture of Authoritarianism in Mexico: A Reexamination." *Latin American Research Review* 19(1): 106–124.

Bourdieu, Pierre. 1969. "Intellectual Field and Creative Project." *Social Science Information* 8:189–219.

———. 1977. *Outline of a Theory of Practice*. Cambridge: Cambridge University Press.

———. 1983. "The Field of Cultural Production, or: The Economic World Reversed." *Poetics* 12(4): 311–356.

———. 1984. *Distinction: A Social Critique of the Judgement of Taste*. Cambridge, MA: Harvard University Press.

———. 1989. "The Field of Power." Unpublished lecture, delivered at the University of Wisconsin at Madison, April.

———. 1990. *The Logic of Practice*. Palo Alto, CA: Stanford University Press.

———. 1993. *The Field of Cultural Production: Essays on Art and Literature*. New York: Columbia University Press.

———. 1998. *The State Nobility: Elite Schools in the Field of Power*. Palo Alto, CA: Stanford University Press.

Bourdieu, Pierre, and Loic Wacquant. 1992. *An Invitation to Reflexive Sociology*. Chicago: University of Chicago Press.

Brain, Stephen. 2011. *Song of the Forest: Russian Forestry and Stalinist Environmentalism, 1905–1953*. Pittsburgh: University of Pittsburgh Press.

Breyfogle, Nicholas. 2015. "At the Watershed: 1958 and the Beginnings of Lake Baikal Environmentalism." *Slavonic and East European Review* 93(1): 147–180.

Breyfogle, Nicholas B. 2013. "The Fate of Fishing in Tsarist Russia: The Human-Fish Nexus in Lake Baikal." *Sibirica* 12(2): 1–29.

Broderson, Arvid. 1957. "National Character: An Old Problem Re-examined." *Diogenes* 20: 84–102.

Brodsky, Joseph. 1986. "In a Room and a Half." *The New York Review of Books*, February 27.

Brown, Kate Pride. 2016. "The Prospectus of Activism: Discerning and Delimiting Imagined Possibility." *Social Movement Studies* 15(6): 547–560.

Bruno, Andy. 2016. *The Nature of Soviet Power*. New York: Cambridge University Press.

Bryanski, Gleb. 2012. "Russia's Putin Signs Anti-Protest Law before Rally." *Reuters*, June 8.

Bull, Bartle. 2001. *Around the Sacred Sea: Mongolia and Lake Baikal on Horseback*. Edinburgh, UK: Canongate.

Burawoy, Michael et al. 2000. *Global Ethnography*. Berkeley: University of California Press.

Businessweek. 2013. "Company Overview of En+ Group Ltd." Accessed on September 5, 2013. http://investing.businessweek.com/research/stocks/private/snapshot.asp?privcapid= 34947186.

Buttel, F. H. 2000. "Ecological Modernization as Social Theory." *Geoforum* 31(1): 57–65.

Carey, Pete. 2013. "Lake Tahoe-Truckee Home Sales Surge Thanks to Bay Area Buyers." *San Jose Mercury News*. Accessed on September 8, 2014. http://www.mercurynews.com/ business/ci_23833936/lake-tahoe-truckee-home-sales-surge-thanks-bay.

Carroll, Archie B., and Kareem Shabana. 2009. "The Business Case for Corporate Social Responsibility: A Review of Concepts, Research and Practice." *International Journal of Management Reviews* 12(1): 85–105.

Chan, Sewell. 2017. "Russia's 'Gay Propaganda' Laws Are Illegal, European Court Rules." *New York Times*, June 20.

Chase, William. 1989. "Voluntarism, Mobilisation and Coercion: Subbotniki 1919–1921." *Europe-Asia Studies* 41(1): 111–128.

Cheloukhine, Serguei, Sanja Kutnjak Ivković, Qasim Haq, and Maria R. Haberfeld. 2015. "Police Integrity in Russia." In *Measuring Police Integrity Across the World*, edited by Sanja Kutnjak Ivković and Maria Haberfeld, 153–181. New York: Springer.

Clark, John D. 2003 *Worlds Apart: Civil Society and the Battle for Ethical Globalization*. New York: Routledge.

Clarke, Simon. 2002. *Making Ends Meet in Contemporary Russia: Secondary Employment, Subsidiary Agriculture and Social Networks*. Cheltenham, UK: Edward Elgar.

Clemens, Elizabeth. 1996. "Organizational Form as Frame: Collective Identity and Political Strategy in the American Labor Movement, 1880–1920." In *Comparative Perspectives on Social Movements: Political Opportunities, Mobilizing Structures, and Cultural Framings*, edited by Doug McAdam, John D. McCarthy, and Mayer Zald, 205–226. New York: Cambridge University Press.

Cohen, Jean, and Andrew Arato. 1993. *Civil Society and Political Theory*. Cambridge, MA: MIT Press.

Colas, Alejandro. 2002. *International Civil Society: Social Movements in World Politics*. Malden, MA: Polity.

Comaroff, John, and Jean Comaroff. 1992. *Ethnography and the Historical Imagination*. Chicago: University of Chicago Press.

Credit Suisse. 2013. "Global Wealth Report." Credit Suisse Research Institute. Accessed on February 1, 2015. https://publications.credit-suisse.com/tasks/render/file/ ?fileID=BCDB1364-A105-0560-1332EC9100FF5C83.

Crotty, Jo, Sarah Marie Hall, and Sergej Ljubownikow. 2014. "Post-Soviet Civil Society Development in the Russian Federation: The Impact of the NGO Law." *Europe-Asia Studies* 66(8): 1253–1269.

Dalton, Russell, Paula Garb, Nicholas Lovrich, John Pierce, and John Whiteley. 1999. *Critical Masses: Citizens, Nuclear Weapons Production, and Environmental Destruction in the United States and Russia.* Cambridge, MA: MIT Press.

Darst, Robert. 2001. *Smokestack Diplomacy.* Cambridge, MA: MIT Press.

Dawisha, Karen. 2015. *Putin's Kleptocracy: Who Owns Russia?* New York: Simon and Schuster.

Dawson, Jane. 1996. *Eco-nationalism: Anti-nuclear Activism and National Identity in Russia, Lithuania, and Ukraine.* Durham, NC: Duke University Press.

Desai, Padma. 2001. *Work without Wages: Russia's Non-Payment Crisis.* Boston: MIT Press.

Diamond, Larry Jay. 1994. "Toward Democratic Consolidation." *Journal of Democracy* 5(3): 4–17.

DiMaggio, Paul, and Walter W. Powell. 1983. "The Iron Cage Revisited: Institutional Isomorphism and Collective Rationality in Organizational Fields." *American Sociological Review* 48: 147–60.

Diment, Galya, and Yuri Slezkine, eds. 1993. *Between Heaven and Hell: The Myth of Siberia in Russian Culture.* New York: Springer.

Dobbin, Frank, Beth Simmons, and Geoffrey Garrett. 2007. "The Global Diffusion of Public Policies: Social Construction, Coercion, Competition, or Learning?" *Annual Review of Sociology* 33: 449–472.

Domrin, Alexander N. 2003. "Ten Years Later: Society, "Civil Society," and the Russian State." *The Russian Review* 62(2): 193–211.

Dore, Ronald, William Lazonick, and Mary O'Sullivan. 1999. "Varieties of Capitalism in the Twentieth Century." *Oxford Review of Economic Policy* 15(4): 102–120.

Dowie, Mark. 1996. *Losing Ground.* Boston: MIT Press.

Dunlop, John B. 1993. *The Rise of Russia and the Fall of the Soviet Empire.* Princeton, NJ: Princeton University Press.

Earle, Richard. 2002. *Art of Cause Marketing.* New York: McGraw-Hill.

Eberly, Don. 2008. *The Rise of Global Civil Society: Building Communities and Nations from the Bottom Up.* New York: Encounter Books.

Elder, Miriam. 2011. "Vladimir Putin Mocks Moscow's 'Condom-Wearing' Protesters." *The Guardian,* December 15.

———. 2012. "Russians Fear Crackdown as Hundreds Are Arrested after Anti-Putin Protest." *The Guardian,* March 6.

Ellen, Pam Scholder, Lois A. Mohr, and Deborah J. Webb. 2000. "Charitable Programs and the Retailer: Do They mix?" *Journal of Retailing* 76(3): 393–406.

Elyachar, Julia. 2006. "Best Practices: Research, Finance, and NGOs in Cairo." *The American Ethnologist* 33(3): 413–426.

En+ Group. 2013. "Strategy." En+ Website. Accessed on September 5, 2013. http://eng.enplus.ru/about/strategy/.

Englund, Will. "Russian Anti-Pollution Activist Pays High Price." *The Washington Post,* July 13.

Enikolopov, Ruben, Vasily Korovkin, Maria Petrova, Konstantin Sonin, and Alexei Zakharov. 2013. "Field Experiment Estimate of Electoral Fraud in Russian Parliamentary Elections." *Proceedings of the National Academy of Sciences* 110(2): 448–452.

European Commission. 2014. "Europe 2020 Civil Society." Accessed on September 9, 2014. http://ec.europa.eu/europe2020/who-does-what/stakeholders/index_en.htm.

Evans, Alfred B.Jr 2002. "Recent Assessments of Social Organizations in Russia." *Demokratizatsiya* 10(3): 322–342.

Evers, Adalbert, and Jean-Louis Laville, eds. 2004. *The Third Sector in Europe.* Northampton, MA: Edward Elgar.

Farrell, Justin. 2016. "Corporate funding and ideological polarization about climate change." *Proceedings of the National Academy of Sciences* 113(1): 92–97.

Federalnii zakon ot 20.07.2012 g. N 121-F3 "O vnesenii izmenenii v otdel'nie zakonodatel'nie akti Rossiiskoi Federatsii v chasti regulirovaniya deyatel'nosti nekommercheskikh organizatii vypolhyayuschikh funktsii inostrannovo agenta."

Feshbach, Murray, and Alfred Friendly. 1992. *Ecocide in the USSR: Health and Nature under Siege*. New York: Basic Books.

Field, Daniel. 1987. "Peasants and Propagandists in the Russian Movement to the People of 1874." *The Journal of Modern History* 59(3): 416–438.

Figes, Orlando. 1996. *A People's Tragedy: The Russian Revolution, 1891–1924*. New York: Penguin Books.

Fligstein, Neil. 2001. "Social Skill and the Theory of Fields." *Sociological Theory* 19(2): 105–125.

Fligstein, Neil, and Doug McAdam. 2012. *A Theory of Fields*. New York: Oxford University Press.

Florini, Ann, ed. 2000. *The Third Force: The Rise of Transnational Civil Society*. Washington, DC: The Carnegie Endowment for International Peace.

Foster, John Bellamy. 2002. *Ecology against Capitalism*. New York: Monthly Review Press.

Foucault, Michel. 1978. *The History of Sexuality, Vol 1*. New York: Vintage.

———. 2003. *Society Must Be Defended: Lectures at the Collège de France 1975–1976*. London: Picador.

———. 2007. *Security, Territory, Population: Lectures at the Collège de France 1977–1978*. London: Picador.

Friedman, Milton. 2002. *Capitalism and Freedom*. Chicago: University of Chicago Press.

Friedman, Elisabeth Jay, Kathryn Hochstetler, and Ann Marie Clark. 2005. *Sovereignty, Democracy, and Global Civil Society*. Albany: SUNY Press.

Frye, Timothy. 2002. "Private Protection in Russia and Poland." *American Journal of Political Science* 46(3): 572–584.

Fukuyama, Francis. 1992. *The End of History and the Last Man*. New York: Free Press.

Galazii, Grigori. 2012. *Baikal v voprosakh i otvetakh*. Saint Petersburg: Forvard.

Garrels, Anne. 2016. *Putin Country: A Journey into the Real Russia*. New York: Farrar, Straus, Giroux.

Gardham, Duncan. 2012. "Berezovsky v Abramovich: How Roman Abramovich Made His Fortune." *The Telegraph*, August 31.

Gellner, Ernest. 1994. *Conditions of Liberty: Civil Society and Its Rivals*. London: Hamish Hamilton.

Gelman, Vladimir. 2015. *Authoritarian Russia: Analyzing Post-Soviet Regime Changes*. Pittsburgh: University of Pittsburgh Press.

Gerth, H. H. and C. Wright Mills, eds. and trans. 1958. *From Max Weber: Essays in Sociology*. New York: Oxford University Press.

Gessen, Masha. 2012. *The Man without a Face*. New York: Riverhead Books.

Geyer, Michael, and Sheila Fitzpatrick. 2008. *Beyond Totalitarianism*. New York: Cambridge University Press.

Gille, Zsuzsa. 2007. *From the Cult of Waste to the Trash Heap of History*. Bloomington: Indiana University Press.

Gille, Zsuzsa, and Sean Riain. 2002. "Global Ethnography." *Annual Review of Sociology* 28: 271–295.

Go, Julian. 2008. "Global Fields and Imperial Forms: Field Theory and the British and American Empires." *Sociological Theory* 26(3): 201–229.

Goffman, Erving. 1959. *The Presentation of Self in Everyday Life*. New York: Anchor Books.

Goldman, Michael. 2001. "Constructing an Environmental State: Eco-governmentality and Other Transnational Practices of a 'Green' World Bank." *Social Problems* 48(4): 499–523.

———. 2007. "How 'Water for All!' Policy Became Hegemonic: The Power of the World Bank and Its Transnational Policy Networks." *GeoForum* 38: 786–800.

Goodwin, Robin, and Peter Allen. 2006. "Democracy and Fatalism in the Former Soviet Union." *Journal of Applied Social Psychology* 30(12): 2558–2574.

Gordon, Jeffrey N. 1999. "Pathways to Corporate Convergence? Two Steps on the Road to Shareholder Capitalism in Germany: Deutsche Telekom and Daimler Chrysler." *Columbia Journal of European Law* 5(2): 219–241.

Gorodnichenko, Yuriy, and Yegor Grygorenko. 2008. "Are Oligarchs Productive? Theory and Evidence." *Journal of Comparative Economics* 36(1): 17–42.

Gorodnichenko, Yuriy, Jorge Martinez-Vazquez, and Klara Sabirianova Peter. 2008. Myth and Reality of Flat Tax Reform: Micro Estimates of Tax Evasion Response and Welfare Effects in Russia. No. w13719. National Bureau of Economic Research.

Graham, Patricia Albjerg. 2005. *Schooling America: How the Public Schools Meet the Nation's Changing Needs.* New York: Oxford University Press.

Green, Eric. 1990. *Ecology and Perestroika: Environmental Protection in the Soviet Union.* A report prepared for the American Committee on US-Soviet Relations.

Guriev, Sergei, and Andrei Rachinsky. 2005. "The Role of Oligarchs in Russian Capitalism." *The Journal of Economic Perspectives* 19(1): 131–150.

Habermas, Jurgen. 1985. *The Theory of Communicative Action, Vol. 2.* Boston, MA: Beacon Press.

Hamm, Patrick, Lawrence P. King, and David Stuckler. 2012. "Mass Privatization, State Capacity, and Economic Growth in Post-communist Countries." *American Sociological Review* 77(2): 295–324.

Harding, Luke. 2010. "Anger at Putin Decision to Allow Lake Baikal Paper Mill to Reopen." *The Guardian*, January 20.

———. 2013. "Boris Berezovsky: A Tale of Revenge, Betrayal and Feuds with Putin." *The Guardian*, March 23.

Havel, Vaclav. 1978. "The Power of the Powerless." Accessed on February 5, 2015. http://vaclavhavel.cz/showtrans.php?cat=eseje&val=2_aj_eseje.html&typ=HTML.

Haydu, Jeffrey. 2008. *Citizen Employers.* Ithaca, NY: Cornell University Press.

Hayek, F. A. 2007. *The Road to Serfdom.* Chicago: University of Chicago Press.

Headley, Jim. 2005. "War On Terror Or Pretext For Power?: Putin, Chechnya, And The 'Terrorist International'." *Australasian Journal of Human Security* 1(2): 13–35.

Hellbeck, Jochen. 2006. *Revolution on My Mind.* Cambridge, MA: Harvard University Press.

Hemment, Julie. 2015. *Youth Politics in Putin's Russia.* Bloomington: Indiana University Press.

Henderson, Sarah. 2003. *Building Democracy in Contemporary Russia: Western Support for Grassroots Organizations.* Ithaca, NY: Cornell University Press.

Hendley, Kathryn. 2012. "Who Are the Legal Nihilists in Russia?" *Post-Soviet Affairs* 28(2): 149–186.

Henry, Laura. 2006. "Shaping Social Activism in Post-Soviet Russia: Leadership, Organizational Diversity, and Innovation." *Post-Soviet Affairs* 22(2): 99–124.

———. 2010. *Red to Green: Environmental Activism in Post-Soviet Russia.* Ithaca, NY: Cornell University Press.

Henry, Laura, and Vladimir Douhovnikoff. 2008. "Environmental Issues in Russia." *Annual Review of Environment and Resources* 33: 437–460.

Henry, Laura A., and Lisa McIntosh Sundstrom. 2007. "Russia and the Kyoto Protocol: Seeking an Alignment of Interests and Image." *Global Environmental Politics* 7(4): 47–69.

Hoff, Karla, and Joseph E. Stiglitz. 2004. "After the Big Bang? Obstacles to the Emergence of the Rule of Law in Post-Communist Societies." *American Economic Review* 94(3): 753–763.

Hosking, Geoffrey. 1991. *The Awakening of the Soviet Union.* Cambridge, MA: Harvard University Press.

———. 2001. *Russia and the Russians: A History.* Cambridge, MA: Belknap.

Howard, Marc Morje. 2003. *The Weakness of Civil Society in Post-Communist Europe.* New York: Cambridge University Press.

Huber, Joseph. 2000. "Towards Industrial Ecology: Sustainable Development as a Concept of Ecological Modernization." *Journal of Environmental Policy and Planning* 2(4): 269–285.

Hyman, Herbert. 1959. *Political Socialization.* New York: Free Press.

Inglehart, Ronald. 1995. "Public Support for Environmental Protection: Objective Problems and Subjective Values in 43 Societies." *PS: Political Science and Politics* 28(1): 57–72.

Irkutskstat. 2014. "Irkutskskaya Oblast. Kratkii Statisticheskii Spravochnik." Irkutskstat.

Jakobson, L. E. V., and Sergey Sanovich. 2010. "The Changing Models of the Russian Third Sector: Import Substitution Phase." *Journal of Civil Society* 6(3): 279–300.

Josephson, Paul R. 1995. "'Projects of the Century' in Soviet History: Large-Scale Technologies from Lenin to Gorbachev." *Technology and Culture* 36(3): 519–559.

―――. 2002. *Industrialized Nature: Brute Force Technology and the Transformation of the Natural World.* Seattle: Island Press.

―――. 2005. *Red Atom: Russia's Nuclear Power Program from Stalin to Today.* Pittsburgh: University of Pittsburgh.

Josephson, Paul, Nicolai Dronin, Ruben Mnatsakanian, Aleh Cherp, Dmitry Efremenko, and Vladislav Larin. 2013. *An Environmental History of Russia.* New York: Cambridge University Press.

Kaldor, Mary. 2003. *Global Civil Society: An Answer to War.* Malden, MA: Blackwell.

Kaufman, Jason. 2003. *For the Common Good?: American Civic Life and the Golden Age of Fraternity.* New York: Oxford University Press.

Kaylan, Malik. 2014. "Kremlin Values: Putin's Strategic Conservatism." *World Affairs*, May/June 2014.

Keenan, Edward. 1986. "Muscovite Political Folkways." *The Russian Review* 45: 115–181.

Kelner, Shaul. 2012. *Tours That Bind: Diaspora, Pilgrimage, and Israeli Birthright Tourism.* New York: New York University Press.

Kenney, Padraic. 2002. *A Carnival of Revolution.* Princeton, NJ: Princeton University Press.

Kim, Sunhyuk. 2000. *The Politics of Democratization in Korea: The Role of Civil Society.* Pittsburgh: University of Pittsburgh Press.

Klebnikov, Paul. 2001. "Gangster Free Capitalism?" *Forbes*, November 26.

Klein, Naomi. 2014. *This Changes Everything.* New York: Simon and Schuster.

Kochan, Nick. 2003. "The Deal That Made a Russian Oligarch." *The Guardian*, July 5.

Komarov, Boris. 1978. *The Destruction of Nature in the Soviet Union.* White Plains, NY: M. E. Sharpe.

Kornhauser, William. 1959. *The Politics of Mass Society.* New York: Free Press.

Kotkin, Stephen. 2008. *Armageddon Averted: The Soviet Collapse 1970–2000.* New York: Oxford University Press.

Kozyreva, P. M., and A. I. Smirnov. 2015. "Difficulties in the Development of the Russian Courts." *Sociological Research* 54(3): 220–237.

Kramer, Andrew. 2006. "Out of Siberia, a Russian Way to Wealth." *New York Times*, August 20.

Krishna, Aradhna, and Uday Rajan. 2009. "Cause Marketing: Spillover Effects of Cause-Related Products in a Product Portfolio." *Management Science* 55(9): 1469–1485.

Lally, Kathy. 2012. "Russian Police Raid Apartments of Protest Leaders." *The Washington Post*, June 11.

LaPorte, Jody, and Danielle N. Lussier. 2011. "What Is the Leninist Legacy? Assessing Twenty Years of Scholarship." *Slavic Review* 70(3): 637–654.

Laqueur, Walter. 2015. *Putinism: Russia and Its Future with the West.* New York: St. Martin's Press.

Laxer, Gordon, and Sandra Halperin, eds. 2003. *Global Civil Society and Its Limits.* New York: Springer.

Ledeneva, Alena. 1998. *An Economy of Favors: Blat, Networking and Informal Exchange.* Cambridge: Cambridge University Press.

Lewin, Moshe. 1985. *The Making of the Soviet System: Essays in the Social History of Interwar Russia.* New York: Pantheon.

Lipman, Maria. 2009. "Media Manipulation and Political Control in Russia." Report. Chatham House.

Lipset, Seymour Martin. 1960. *Political Man.* New York: Anchor Books.

Liu, Caitlin. 2008. "Domestic Tax Havens." *Upstart Business Journal*, March 31. http://upstart.bizjournals.com/resources/business-intelligence/2008/03/31/Domestic-Tax-Havens.html?page=all.

Loiko, Sergei. 2012. "Putin's Opponents Feel the Heat in Russia." *Los Angeles Times*, August 15.

Long, Lucy, ed. 2004. *Culinary Tourism.* Lexington: University Press of Kentucky.

Madison, James, Alexander Hamilton, and John Jay. 1987. *The Federalist Papers.* New York: Penguin Books.

Makhnyova, Alyona. 2016. "Agenti ekologii." *Vostochno-Sibirskaya Pravda*, February 16.

Margolis, Joshua, and James Walsh. 2003. "Misery Loves Companies: Rethinking Social Initiatives by Business." *Administrative Science Quarterly* 48: 268–305.

Martin, John Levi. 2003. "What Is Field Theory?" American Journal of Sociology 109(1): 1–49.

Marx, Karl. 1964. *Economic and Philosophic Manuscripts of 1844.* New York: International.

———. 1976. *Capital.* New York: International.

Mathiessen, Peter. 1992. *Baikal: Sacred Sea of Siberia.* San Francisco: Sierra Club Books.

Mazour, Anatole G. 1937. *The First Russian Revolution, 1825: The Decembrist Movement, its Origins, Development, and Significance.* Berkeley, CA: University of California Press.

McAdam, Doug. 1986. "Recruitment to High-Risk Activism: The Case of Freedom Summer." *American Journal of Sociology* 92(1): 64–90.

———. 1988. *Freedom Summer.* New York: Oxford University Press.

McCarthy, John, and Mayer Zald. 1977. "Resource Mobilization and Social Movements: A Partial Theory." *The American Journal of Sociology* 82(6): 1212–1241.

McDonnell, Terrence. 2010. "Cultural Objects as Objects: Materiality, Urban Space, and the Interpretation of AIDS Campaigns in Accra, Ghana." *American Journal of Sociology* 115(6): 1800–1852.

McFarland, Daniel, and Reuben Thomas. 2006. "Bowling Young: How Youth Voluntary Associations Influence Adult Political Participation." *American Sociological Review* 71(3): 401–425.

Mendelson, Sarah E. 2008. "Dreaming of a Democratic Russia: A Year in Moscow Promoting a Post-Soviet Political Process." *American Scholar* 77(1): 35.

Merelman, Richard M. 1969. "The Development of Political Ideology: A Framework for the Analysis of Political Socialization." *American Political Science Review* 63(3): 750–767.

Meyer, Henry. 2011. "Russia Repels Retailers as Ikea Halt Curtails Medvedev Goal." *Bloomberg Business,* March 2. http://www.bloomberg.com/news/articles/2011-03-01/russia-repels-retailers-as-ikea-halt-curtails-medvedev-bric-goal.

Meyer, John W., and W. Richard Scott. 1992. *Organizational Environments: Ritual and Rationality.* New York: Sage.

Mill, J. S. 1978. *On Liberty.* Indianapolis: Hackett.

Mishler, William, and Richard Rose. 2005. "What Are the Political Consequences of Trust? A Test of Cultural and Institutional Theories in Russia." *Comparative Political Studies* 38(9): 1050–1078.

Montesquieu, Charles De Secondat. 1989. *The Spirit of the Laws.* Cambridge: Cambridge University Press.

Myers, Steven Lee. 2015. *The New Tsar: The Rise and Reign of Vladimir Putin.* New York: Knopf.

Nan, Xiaoli, and Kwangjun Heo. 2007. "Consumer Responses to Corporate Social Responsibility (CSR) Initiatives: Examining the Role of Brand-Cause Fit in Cause-Related Marketing." *Journal of Advertising* 36(2): 63–74.

Naumov, Igor V. 2006. *The History of Siberia.* New York: Routledge.

Nordhaus, Ted, and Michael Shellenberger. 2007. *Break Through: From the Death of Environmentalism to the Politics of Possibility.* Boston: Houghton Mifflin.

Offord, Derek. 1986. *The Russian Revolutionary Movement in the 1880s.* New York: Cambridge University Press.

Omi, Michael, and Howard Winant. 1986. *Racial Formation in the United States.* New York: Routledge.

Oreskes, Naomi, and Erik Conway. 2011. *Merchants of Doubt.* New York: Bloomsbury.

Paxson, Margaret. 2006. *Solovyovo: The Story of Memory in a Russian Village.* Bloomington: Indiana University Press.

Peck, Tom. 2011. "Abramovich Tells of Role in Aluminium Wars." *The Independent,* November 4.

Perez, Evan, and Gregory L. White. 2009. "FBI Lets Barred Tycoon Visit U.S." *The Wall Street Journal,* October 30.

Peterson, D. J. 1993. *Troubled Lands: The Legacy of Soviet Environmental Destruction.* Boulder, CO: Westview.

Petryna, Adriana. 2002. *Life Exposed: Biological Citizens after Chernobyl*. Princeton, NJ: Princeton University Press.

Pew Research Center. 2012. "Attitudes toward Democracy." In *Russians Back Protests, Political Freedoms and Putin, Too*, 6–21. Global Attitudes Project.

Pipes, Richard. 1964. "Narodnichestvo: A Semantic Inquiry." *Slavic Review* 23(3): 441–458.

Piven, Francis Fox, and Richard Cloward. 1977. *Poor People's Movements: How They Succeed, Why They Fail*. New York: Pantheon.

Pollan, Michael. 2006. *The Omnivore's Dilemma*. New York: Penguin.

Pryde, Philip. 1991. *Environmental Management in the Soviet Union*. Cambridge: Cambridge University Press.

Putnam, Robert D. 1994. "What Makes Democracy Work?" *Review-Institute of Public Affairs* 47(1): 31–34.

———. 2000. *Bowling Alone*. New York: Simon and Schuster.

Rasputin, Valentin. 1996. *Siberia, Siberia*. Evanston, IL: Northwestern University Press.

Reguly, Eric. 2011. "At Home with Russian Oligarch Oleg Deripaska." *The Globe and Mail*, February 11.

Reis, Nancy. 1997. *Russian Talk: Culture and Conversation during Perestroika*. Ithaca, NY: Cornell University Press.

Richter, James, and Walter F. Hatch. 2013. "Organizing Civil Society in Russia and China: A Comparative Approach." *International Journal of Politics, Culture, and Society* 26(4): 323–347.

Roe, Alan. 2016. *Into Soviet Nature: Tourism, Environmental Protection, & the Formation of Soviet National Parks, 1950s-1990s*. PhD Dissertation, Georgetown University.

Russian Financial Control Monitor. 2010. "Deripaska to Sell Baikal Pulp and Paper Mill." *Russian Financial Control Monitor: Mergers and Acquisitions*, July 5.

Sachs, Jeffrey. 2012. "What I Did in Russia." Jeffrey Sachs official website. Accessed on October 23, 2016. http://jeffsachs.org/2012/03/what-i-did-in-russia/.

Sachs, Jeffrey D., and Wing Thye Woo. 1994. "Experiences in the Transition to a Market Economy." *Journal of Comparative Economics* 18(3): 271–275.

Saez, Emmanuel, and Gabriel Zucman. 2014. "Wealth Inequality in the United State Since 1913." NBER Working Paper No. 20625. Cambridge, MA: National Bureau of Economic Research.

Said, Edward. 1978. *Orientalism*. New York: Pantheon.

Sakwa, Richard. 2007. *Putin: Russia's Choice*. New York: Routledge.

Schnaiberg, Allan. 1980. *The Environment: From Surplus to Scarcity*. New York: Oxford University Press.

Schofer, Evan. 2003. "The Global Institutionalization of Geological Science, 1800–1990." *American Sociological Review* 68(Dec.): 730–759.

Schofer, Evan, and Ann Hironaka. 2005. "The Effects of World Society on Environmental Protection Outcomes." *Social Forces* 84(1):25–47.

Schofer, Evan, and John Meyer. 2005. "The World-Wide Expansion of Higher Education in the Twentieth Century." *American Sociological Review* 70: 898–920.

Schussman, Alan, and Sarah Soule. 2005. "Process and Protest: Accounting for Individual Protest Participation." *Social Forces* 84(2): 1083–1108.

Schwirtz, Michael. 2011. "Few at Putin Party's Rally, and Even Fewer Willingly." *New York Times*, December 12.

Schwirtz, Michael, and David Herszenhorn. 2011. "Voters Watch Polls in Russia, and Fraud Is What They See." *New York Times*, December 5.

Scott, James C. 1998. *Seeing Like a State*. New Haven, CT: Yale University Press.

Seidman, Gay. 2007. *Beyond the Boycott*. New York: Russell Sage.

Seligson, Mitchell A., and John A. Booth. 1993. "Political Culture and Regime Type: Evidence from Nicaragua and Costa Rica." *The Journal of Politics* 55(3): 777–792.

Shelley, Louise. 1995. "Privatization and Crime: The Post-Soviet Experience." *Journal of Contemporary Criminal Justice* 11(4): 244–256.

Shevchenko, Olga. 2008. *Crisis and the Everyday in Postsocialist Moscow*. Bloomington, IN: Indiana University Press.

Shevtsova, Lilia. 2007. *Russia–Lost in Transition. The Yeltsin and Putin Legacies*. Washington, DC: Carnegie Endowment for International Peace.

Shklovsky, Victor. 2009. "Art as Device," in *Theory of Prose*, 1–14, translated by Benjamin Sher. London: Dalkey Archive Press.

Shtil'mark, Feliks Robertovich. 2003. *History of the Russian Zapovedniks, 1895–1995*. Devon, UK: Russian Nature Press.

Simpson, Glenn, and Schmidt, Susan. 2008. "Russia's Deripaska Faces Western Investigations." *Wall Street Journal*, October 10

Sixsmith, Martin. 2010. *Putin's Oil: The Yukos Affair and the Struggle for Russia*. London: A and C Black.

Skocpol, Theda. 2003. *Diminished Democracy: From Membership to Management in American Civic Life*. Norman: University of Oklahoma Press.

Smith, Jackie. 2008. *Social Movements for Global Democracy*. Baltimore: Johns Hopkins University Press.

Smith, Scott, and David Alcorn 1991. "Cause Marketing: A New Direction in the Marketing of Corporate Responsibility." *Journal of Consumer Marketing* 8(3): 19–35.

Southworth, Caleb. 2006. "The Dacha Debate: Household Agriculture and Labor Markets in Post-Socialist Russia." *Rural Sociology* 71(3): 451–478.

Sperling, Valerie. 2014. *Sex, Politics, and Putin: Political Legitimacy in Russia*. New York: Oxford University Press.

Speth, James Gustave. 2008. *The Bridge at the Edge of the World: Capitalism, the Environment, and Crossing from Crisis to Sustainability*. New Haven, CT: Yale University Press.

Stanley, Alessandra. 1996. "Russian Banking Scandal Poses Threat to Future of Privatization." *New York Times*, January 28.

Starr, S. Frederick. 1988. "Soviet Union: A Civil Society." *Foreign Policy* 70 (Spring): 26–41.

Steurer, Reinhard. 2013. "Disentangling Governance: A Synoptic View of Regulation by Government, Business and Civil Society." *Policy Sciences* 46(4): 387–410.

Stiglitz, Joseph. 2002. *Globalization and Its Discontents*. New York: Norton.

Swidler, Ann. 1986. "Culture in Action: Symbols and Strategies." *American Sociological Review* 51(2): 273–286.

TASS. 2012. "Court Upholds Decision on Debt Payment Deferral for Baikal TsBK." *TASS*, April 6. http://tass.com/archive/673099.

Thomson, Peter. 2009. *Sacred Sea*. New York: Oxford University Press.

Tiano, Susan. 1986. "Authoritarianism and Political Culture in Argentina and Chile in the Mid-1960's." *Latin American Research Review* 21(1): 73–98.

Tilly, Charles. 1978. *From Mobilization to Revolution*. Boston: Addison-Wesley.

———. 2004. *Social Movements, 1768–2004*. Boulder, CO: Paradigm.

Tocqueville, Alexis de. 1981. *Democracy in America*. New York: Modern Library.

Traynor, Ian. 2004. "US Campaign behind the Turmoil in Kiev." *The Guardian* Thursday, November 25.

Treisman, Daniel. 2010. "'Loans for Shares' Revisited." *Post-Soviet Affairs* 26(3): 207–227.

Trochev, Alexei. 2016. "Can Weak States Have Strong Courts? Evidence from Post-Communist Russia." In *Legitimacy, Legal Development and Change: Law and Modernization Reconsidered*, edited by David K. Linnan, 351–374. New York: Routledge.

Tsutsui, Kiyoteru, and Christine Min Wotipka. 2004. "Global Civil Society and the International Human Rights Movement: Citizen Participation in Human Rights International Nongovernmental Organizations." *Social Forces* 83(2): 587–620.

Tsygankov, Andrei. 2014. *The Strong State in Russia: Development and Crisis*. New York: Oxford University Press.

Turner, Victor. 1974. *Dramas, Fields, and Metaphors*. Ithaca, NY: Cornell University Press.

Turner, Tom. 2015. *David Brower: The Making of the Environmental Movement.* Berkeley: University of California Press.

Varadarajan, P. Rajan, and Anil Menon. 1988. "Cause-Related Marketing: A Coalignment of Marketing Strategy and Corporate Philanthropy." *Journal of Marketing* 52(3): 58–74.

Varese, Federico. 1997. *The Russian Mafia: Private Protection in a New Market Economy.* New York: Oxford University Press.

———. 2004. "Mafia Transplantation." In *Creating Social Trust in Post-Socialist Transition,* 148–166. New York: Palgrave Macmillan.

Viterna, Jocelyn, and Cassandra Robertson. 2015. "New Directions for the Sociology of Development." *Annual Review of Sociology* 41: 243–269.

Viterna, Jocelyn, Emily Clough, and Killian Clarke. 2015. "Reclaiming the 'Third Sector' from 'Civil Society'." *Sociology of Development* 1(1): 173–207.

Walzer, Michael, ed. 1995. *Toward a Global Civil Society,* vol. 1. New York: Berghahn Books.

Ward, Christopher J. 2009. *Brezhnev's Folly: The Building of BAM and Late Soviet Socialism.* Pittsburgh: University of Pittsburgh Press, 2009.

Wareham, Jonathan, and Han Garrits. 1999. "De-contextualising Competence: Can Business Best Practice Be Bundled and Sold?" *European Management Journal* 17(1): 39–49.

Wartick, Steven, and Philip Cochran. 1985. "The Evolution of the Corporate Social Performance Model." *The Academy of Management Review* 10(4): 758–769.

Weber, Christopher, and H. Scott Matthews. 2008. "Food-Miles and the Relative Climate Impacts of Food Choices in the United States." *Environmental Science and Engineering* 42(10): 3508–3513.

Weiner, Douglas. 1988. *Models of Nature.* Pittsburgh: University of Pittsburgh Press.

———. 1999. *A Little Corner of Freedom: Russian Nature Protection from Stalin to Gorbachëv.* Berkeley: University of California Press.

Whitefield, Stephen. 2003. "Russian Mass Attitudes towards the Environment, 1993–2001." *Post-Soviet Affairs* 19(2): 95–113.

Williams, Sean. 2013. "This Is One Incredible CEO." *The Motley Fool,* August 21.

Willis, Paul. 1981. *Learning to Labor: How Working Class Kids Get Working Class Jobs.* New York: Columbia University Press.

World Bank. 2012. *World Income Distribution 2012.* Washington, DC: World Bank.

Wright, Erik Olin. 2010. *Envisioning Real Utopias.* New York: Verso.

Wyss, Robert. 2016. *The Man Who Built the Sierra Club: A Life of David Brower.* New York: Columbia University Press.

Yanitsky, Oleg. 2011. "The Struggle in Defense of Baikal: The Shift of Values and Disposition of Forces." *International Review of Social Research* 1(3): 33–51.

Young, Stephen, ed. 2001. *The Emergence of Ecological Modernization.* London: Routledge.

Yurchak, Alexei. 2006. *Everything Was Forever, Until It Was No More: The Last Soviet Generation.* Princeton, NJ: Princeton University Press.

Zakaria, Fareed. 2014. "The Rise of Putinism." *The Washington Post,* July 31.

Zolotukhina, Elizabeth. 2013. "Why Khodorkovsky?" The Institute of Modern Russia. Accessed on January 31, 2015. http://imrussia.org/en/society/500-why-khodorkovsky

Zygar, Mikhail. 2016. *All the Kremlin's Men: Inside the Court of Vladimir Putin.* New York: Public Affairs.

Index